Microwave and Ultrasound-Assisted Extraction for Plant Bioactive Compounds

Microwave and Ultrasound-Assisted Extraction for Plant Bioactive Compounds

Guest Editors

Ivona Elez Garofulić
Maja Repajić

MDPI

Basel • Beijing • Wuhan • Barcelona • Belgrade • Novi Sad • Cluj • Manchester

Guest Editors
Ivona Elez Garofulić
Department of Food
Engineering
University of Zagreb
Zagreb
Croatia

Maja Repajić
Department of Food
Engineering
University of Zagreb
Zagreb
Croatia

Editorial Office
MDPI AG
Grosspeteranlage 5
4052 Basel, Switzerland

This is a reprint of the Special Issue, published open access by the journal *Molecules* (ISSN 1420-3049), freely accessible at: https://www.mdpi.com/journal/molecules/special_issues/9BQ2GJ09DB.

For citation purposes, cite each article independently as indicated on the article page online and as indicated below:

Lastname, A.A.; Lastname, B.B. Article Title. *Journal Name* **Year**, *Volume Number*, Page Range.

ISBN 978-3-7258-8019-5 (Hbk)
ISBN 978-3-7258-8020-1 (PDF)
https://doi.org/10.3390/books978-3-7258-8020-1

Cover image courtesy of Josipa Dukić
High-Power Ultrasound and High-Voltage Electrical Discharge-Assisted Extractions of Bioactive Compounds

Contents

molecules

MDPI

Article

Multi-Response Modeling for Bio-Compound Ultrasound-Assisted Extraction (UAE) from Matico (*Piper aduncum* L.) and Chacruna (*Psychotria viridis* Ruiz & Pav.) Leaves Originating in the Peruvian Amazon

Raquel Rafael-Saldaña [1,†], Roifer Pérez-Vasquez [1,†], José Luis Pasquel-Reátegui [1,*], Manuel Fernando Coronado-Jorge [1], Pierre Vidaurre-Rojas [2], Ángel Cárdenas-García [3], Keller Sánchez-Dávila [4] and Keneth Reátegui-Del Águila [5]

1 Grupo de Investigación en Ingeniería y Tecnología Agroindustrial, Facultad de Ingeniería Agroindustrial, Universidad Nacional de San Martín (UNSM), Tarapoto 22000, Peru; rrafaels@alumno.unsm.edu.pe (R.R.-S.); rperezv@alumno.unsm.edu.pe (R.P.-V.); mfcoronado@unsm.edu.pe (M.F.C.-J.)

2 Grupo de Investigación Gestión ATEC, Facultad de Ciencias Económicas, Universidad Nacional de San Martín (UNSM), Tarapoto 22200, Peru; pvidaurre@unsm.edu.pe

3 Facultad de Ingeniería de Sistemas e Informática, Universidad Nacional de San Martín (UNSM), Tarapoto 22201, Peru; acardenasg@unsm.edu.pe

4 Facultad de Medicina Humana, Universidad Nacional de San Martín (UNSM), Tarapoto 22000, Peru; ksanchezd@unsm.edu.pe

5 Faculty of Engineering and Environmental Sciences, Universidad Nacional Intercultural de la Amazonia (UNIA), Pucallpa 22200, Peru; kreateguid@unia.edu.pe

* Correspondence: jlpasquelr@unsm.edu.pe; Tel.: +51-949132754

† These authors contributed equally to this work.

Abstract: Medicinal plants play an essential role in the food, pharmaceutical, and cosmetic industries due to their ability to prevent and treat diseases. In this study, a three-factor, three-level Box–Behnken experimental design (BBD) with response surface methodology (RSM) was used to maximize the conditions of ultrasound-assisted extraction (UAE) of bioactive compounds from matico and chacruna leaves in terms of total extraction yield (TEY), total phenolic content (TPC) and antioxidant activity (AA) using ABTS and DPPH assays. The effect of methanol concentration (X1: 25%, 50%, and 75%), time (X2: 3, 6, and 9 min), and power (X3: 90, 270, and 450 W) was evaluated as independent variables. The experimental results were fitted to second-order polynomial models, and multiple regression analysis and analysis of variance were used to determine the suitability of the models, using which 3D response surface plots were generated. Considering the multivariable optimization, the best extraction conditions were 73.68% *v*/*v* methanol, 9 min, 269.32 W for matico, and 64.84% *v*/*v* methanol, 3 min, 344.44 W for chacruna. Under these conditions, the maximum value of 18.33 and 20.83% for TEY, 7.16 and 40.86 mg GAE/g dm for TPC, 56.88 and 526.38 µmol TE/g dm for DPPH were predicted for matico and chacruna, respectively. Practical Applications: This research focused on the modeling by response surface methodology (RSM) of Ultrasound-Assisted Extraction of bioactive compounds from matico and chacruna, Peruvian plants used in traditional medicine. The methodologies used allow the maximization of bioactive extraction, which presented a high recovery of phenolics with high antioxidant activity. These results highlight the use of Amazon plants in traditional medicine and their possible use in other industries such as cosmetic or food safety.

Keywords: response surface; yield; phenolics; antioxidants; Box–Behnken; medicinal plants

1. Introduction

Several plant species have a fairly widespread empirical use due to their medicinal properties [1]. Medicinal plants are a more affordable and convenient alternative for treating multiple conditions and are linked to their active constituents [2]. In addition, many drugs currently used were developed from plants used in traditional medicine, so there is great interest in the rediscovery of medicinal plants as a source of possible medicines [3].

The active principles identified in various plant species can be used to formulate medicines, cosmetics, and food products [4]. Although synthetic molecules with biological activity are currently available, consumers prefer products of natural origin because they are perceived as safer, and that is why plants continue to be used for medicinal purposes [5].

Piper aduncum L. (matico) is a species native to Peru and is frequently used in traditional medicine [6]; antiviral, antifungal, and antibacterial properties are attributed to it, and it is used for the treatment of nausea, ulcers, hemorrhages, cystitis, respiratory diseases, wounds, and muscle injuries, which is supported by the presence of secondary metabolites found in the phytochemical analysis [7].

Psychotria viridis (chacruna) is used to treat fever and is a key component of the famous drink "ayahuasca". The plant contains N,N-dimethyltryptamine, which, when combined with monoamine oxidase inhibitors, produces psychoactive and hallucinogenic effects [8]. Beyond its traditional use, *Psychotria viridis* has medicinal applications as an antidepressant, anticancer, and antiviral agent [9]. The majority of studies focus on the determination and quantification of N,N-dimethyltryptamine (DMT) [10].

The extraction of natural phenolic compounds has been the subject of interest due to their high antioxidant activity and applicability in the food, medicinal, and pharmaceutical fields [11]. Extraction is the first and most important step in recovering and purifying bioactive compounds from plants [12]. Conventional techniques have disadvantages, such as long times, residual solvents, high costs, low flexibility, and degradation of the compounds of interest, and others [13].

Ultrasound-assisted extraction (UAE) is a novel alternative considered a green technology, with advantages such as short extraction times, low temperature, lower energy, and solvent consumption, in addition to having scalability potential [14]. However, it becomes more efficient due to the phenomenon of acoustic cavitation, which originates from the creation, growth, and implosion of gas bubbles that collapse on the surface of the plant material, releasing high pressure and temperature, thereby facilitating the penetration of the solvent [15].

The response surface methodology (RSM) is a statistical and mathematical technique that allows modeling and optimizing extraction processes of bioactive components [16], through a polynomial equation, a model that relates the effect of the independent variables and their interaction on the response variable [17], and also reduces the number of experiments compared to a complete factorial design [18]. The Box–Behnken design (BBD) is one of the most used in RSM [19,20].

Matico and chacruna are frequently used in traditional Peruvian medicine, but there needs to be more scientific evidence to validate their clinical use. The vast majority of published works deal with extracts obtained by maceration and isolated compounds, especially in the field of chemistry and ethnopharmacology [1,21]. Therefore, the objective of this work was to optimize the ultrasound-assisted extraction (UAE) of matico and chacruna leaves by applying a Box–Behnken design with RSM, in order to evaluate the effect of the extraction conditions: methanol concentration (X1: 25, 50 and 75%), time (X2: 3, 6 and 9 min) and power (X3: 90, 270 and 450 W), on the total extraction yield (TEY), total phenolic content (TPC), and antioxidant activity (AA) by ABTS and DPPH assays.

2. Results

2.1. Characterization of Matico and Chacruna Leaves

The crushed dried leaves of *Piper modicum* L. and *Psychotria viridis* Ruiz & Pav. were characterized in terms of crude fiber, fat, moisture, total protein, total ash, carbohydrates, and total energy (Table 1).

Table 1. Proximal composition of matico (*Piper aducum* L.) and chacruna (*Psychotria viridis* Ruiz & Pav.) leaves.

	***Piper aducum* L.**	***Psychotria viridis* Ruiz & Pav.**
Moisture (%)	6.7 ± 0.11	10.5 ± 0.02
Total protein (%)	20.8 ± 0.00	10.3 ± 0.06
Fat (%)	4.9 ± 0.00	0.7 ± 0.00
Total ash (%)	14.2 ± 0.01	10.4 ± 0.01
Carbohydrate (%)	53.4	68.1
Crude fiber (%)	10.7 ± 0.05	16.2 ± 0.04
Total Energy (kcal)	340.9	319.9

Note: The carbohydrate content was determined by difference (fats, moisture, proteins, and ashes).

2.2. Using Box–Behnken Experimental Design

The Box–Behnken experimental design was used to evaluate the effect of variables on the extraction process. Tables 2 and 3 present the experimental and predicted data obtained under the conditions established by the experimental design for matico and chacruna, respectively. Response surface graphs were constructed to visualize the effect of the response variables on the extraction of the compounds of interest. Two variables were varied over different ranges, while the third variable was kept constant [22].

Table 2. Box–Behnken design matrix and experimental results obtained in the ultrasound-assisted extraction of bioactive compounds from matico.

E	Independent Variables			Chemical Characterization of the Extracts							
	X_1 (%)	X_2 (min)	X_3 (W)	TEY	A	TPC	A	ABTS	A	DPPH	A
01	−1 (25)	−1 (3)	0 (270)	20.52 ± 1.06	20.15	3.38 ± 0.21	3.32	48.49 ± 1.94	48.84	16.19 ± 0.5	13.35
02	+1 (75)	−1 (3)	0 (270)	14.97 ± 0.34	15.72	5.95 ± 0.59	5.98	85.7 ± 4.7	85.33	45.45 ± 0.64	51.92
03	−1 (25)	+1 (9)	0 (270)	21.58 ± 0.41	22.35	2.95 ± 0.16	2.90	41.2 ± 3.48	42.82	15.23 ± 1.36	19.84
04	+1 (75)	+1 (9)	0 (270)	18.7 ± 0.16	17.91	7.27 ± 0.18	7.32	92.95 ± 2.43	93.85	57.27 ± 2.71	58.41
05	−1 (25)	0 (6)	−1 (90)	21.41 ± 0.3	21.25	3.08 ± 0.31	3.11	50.12 ± 4.91	45.83	16.97 ± 0.23	16.60
06	+1 (75)	0 (6)	−1 (90)	15.67 ± 0.37	16.01	6.3 ± 0.1	6.65	88.91 ± 7.36	89.59	54.86 ± 4.95	55.17
07	−1 (25)	0 (6)	+1 (450)	21.49 ± 0.96	21.25	3.03 ± 0.28	3.11	43.51 ± 0.51	45.83	17.99 ± 0.61	16.60
08	+1 (75)	0 (6)	+1 (450)	17.91 ± 1.06	16.81	7.08 ± 0.21	6.65	90.81 ± 5.39	89.59	63.09 ± 2.12	55.17
09	0 (50)	−1 (3)	−1 (90)	20.41 ± 1.58	20.92	4.08 ± 0.37	4.21	54.79 ± 3.16	55.09	28.02 ± 0.56	27.62
10	0 (50)	+1 (9)	−1 (90)	21.62 ± 0.67	23.11	4.58 ± 0.11	4.67	58.04 ± 4.98	56.34	40.9 ± 3.19	34.11
11	0 (50)	−1 (3)	+1 (450)	21.84 ± 0.28	20.92	4.54 ± 0.36	4.21	59.33 ± 1.89	55.09	29.18 ± 1.07	27.62
12	0 (50)	+1 (9)	+1 (450)	24.61 ± 0.41	23.11	4.99 ± 0.43	4.67	61.13 ± 1.91	56.34	31.4 ± 2.8	34.11
13	0 (50)	0 (6)	0 (270)	22.45 ± 0.48	22.02	4.59 ± 0.09	4.44	57.07 ± 4.5	55.72	30.21 ± 0.83	30.86
14	0 (50)	0 (6)	0 (270)	21.12 ± 0.54	22.02	3.96 ± 0.35	4.44	52.86 ± 4.27	55.72	28.94 ± 0.17	30.86
15	0 (50)	0 (6)	0 (270)	22.08 ± 0.16	22.02	4.37 ± 0.08	4.44	46.79 ± 4.26	55.72	27.38 ± 1.81	30.86

Data are shown as means ± standard deviations (n = 3). E = experiment; X_1 = methanol concentration; X_2 = time; X_3 = power; A = predicted; TEY = total extraction yield (%); TPC = total phenolic content (mg GAE/g dm); ABTS = antioxidant activity by ABTS assay (µmol TE/g dm); DPPH = antioxidant activity by DPPH assay (µmol TE/g dm).

Table 3. Box–Behnken design matrix and experimental results obtained in the ultrasound-assisted extraction of bioactive compounds from chacruna.

E	Independent Variables			Chemical Characterization of the Extracts							
	X_1 (%)	X_2 (min)	X_3 (W)	TEY	A	TPC	A	ABTS	A	DPPH	A
01	−1 (25)	−1 (3)	0 (270)	17.3 ± 0.4	18.15	30.17 ± 1.22	29.65	562.95 ± 22.39	560.78	365.51 ± 33.98	362.11
02	+1 (75)	−1 (3)	0 (270)	21.2 ± 0.35	18.86	45.73 ± 1.03	40.80	939.72 ± 33.85	822.93	547.55 ± 46.16	500.27

Table 3. *Cont.*

E	Independent Variables			Chemical Characterization of the Extracts							
	X_1 (%)	X_2 (min)	X_3 (W)	TEY	A	TPC	A	ABTS	A	DPPH	A
03	−1 (25)	+1 (9)	0 (270)	15.39 ± 0.82	17.73	22.6 ± 1.13	27.53	448.82 ± 3.77	565.61	298.85 ± 7.42	346.13
04	+1 (75)	+1 (9)	0 (270)	17.86 ± 0.42	17.01	31.83 ± 1.04	32.35	591.54 ± 4.56	593.72	404.76 ± 22.14	408.16
05	−1 (25)	0 (6)	−1 (90)	15.8 ± 0.71	14.19	21.6 ± 0.83	19.57	457.95 ± 9.44	409.71	297.24 ± 9.11	283.38
06	+1 (75)	0 (6)	−1 (90)	9.43 ± 0.18	11.01	20.75 ± 1.16	23.13	422.41 ± 19.92	488.79	276.08 ± 23.97	306.10
07	−1 (25)	0 (6)	+1 (450)	14.86 ± 0.11	13.28	19.66 ± 1.69	17.28	438.11 ± 31.37	371.73	285.05 ± 7.4	255.04
08	+1 (75)	0 (6)	+1 (450)	14.84 ± 0.31	16.45	27.67 ± 1.31	29.70	534.68 ± 3.94	582.92	418.65 ± 23.27	432.51
09	0 (50)	−1 (3)	−1 (90)	16.21 ± 0.45	16.97	29.41 ± 0.65	31.96	543.02 ± 16.33	593.43	371.41 ± 38.09	388.67
10	0 (50)	+1 (9)	−1 (90)	16.96 ± 0.57	16.23	29.48 ± 0.56	26.58	554.51 ± 9.94	485.96	412.7 ± 35.48	379.29
11	0 (50)	−1 (3)	+1 (450)	18.88 ± 0.43	19.62	31.09 ± 0.37	33.99	557.67 ± 10.29	626.22	448.96 ± 41.01	482.37
12	0 (50)	+1 (9)	+1 (450)	18.87 ± 1.15	18.11	31.36 ± 0.98	28.81	559.73 ± 14.73	509.32	400.92 ± 8.15	383.66
13	0 (50)	0 (6)	0 (270)	20.14 ± 0.03	19.06	33.14 ± 0.99	32.01	587.78 ± 17.88	575.36	470.4 ± 37.84	418.00
14	0 (50)	0 (6)	0 (270)	18.66 ± 0.81	19.06	32.61 ± 1.63	32.01	572.25 ± 15.76	575.36	426.59 ± 29.53	418.00
15	0 (50)	0 (6)	0 (270)	18.39 ± 0.76	19.06	30.29 ± 0.75	32.01	566.05 ± 18.12	575.36	357 ± 22.33	418.00

Data are shown as means ± standard deviations (n = 3). E = experiment; X_1 = methanol concentration; X_2 = time; X_3 = power; A = predicted; TEY = total extraction yield (%); TPC = total phenolic content (mg GAE/g dm); ABTS = antioxidant activity by ABTS assay (µmol TE/g dm); DPPH = antioxidant activity by DPPH assay (µmol TE/g dm).

2.2.1. UAE of Matico (*Piper aduncum* L.) Leaves

Methanol concentration had the most significant effect on TEY, TPC, ABTS, and DPPH extraction from matico leaves, followed by time, while power had no effect on the responses (Tables S1–S4).

The effects of the variables on TPC, ABTS, and DPPH were very similar. The high impact of the methanol concentration can be observed in experiments 3 and 4, where the values increased when varying the methanol concentration from 25% to 75%, reaching 2.95 and 7.27 mg GAE/g dm, 41.2 and 92.95 µmol TE/g dm, 15.23 and 57.27 µmol TE/g dm, respectively, for TPC, ABTS, and DPPH (Table 2).

Conversely, the highest TEY values were achieved with low methanol concentrations. For example, in experiments 1 and 2, 20.52% and 14.97% were obtained using 25% and 75% methanol, respectively.

Experiments 10 and 11 illustrate the low influence of time and the absence of the power effect on TPC and ABTS. Similar results were achieved by combining 9 min with 90 W and 3 min with 450 W (4.58 ± 0.11 and 4.54 ± 0.36 mg GAE/g dm; 58.04 ± 4.98 and 59.33 ± 1.89 µmol TE/g dm, for TPC and ABTS, respectively). The effect of time was more significant for the values of TEY and DPPH. This behavior can be observed in experiments 2 and 4 for TEY (14.97% and 18.7%) and in experiments 9 and 10 for DPPH (28.02 and 40.9 µmol TE/g dm), with times of 3 min and 9 min, respectively, under the same methanol concentration and power conditions.

The proposed models have demonstrated their reliability by adequately explaining the experimental data. The lack of adjustment was not significant, affirming their suitability (Table 4). The optimization at the levels studied predicts a maximum value of 23.53% using 40.71% methanol, 9 min, and 173.3 W for TEY. It predicts a result of 7.32 mg GAE/g dm, 93.85 µmol TE/g dm, and 58.41 µmol TE/g dm for TPC, ABTS and DPPH, respectively, under the conditions of 75% methanol, 9 min, and 270 W.

Figure 1a shows a small optimized area for TEY, with a methanol concentration between 35% and 45%. Based on this, the matico leaves appear to contain more polar compounds than nonpolar ones. The response surface plots for TPC, ABTS, and DPPH did not display optimal regions under the applied conditions (Figure 1b–d). These results suggest that the extraction process may be further improved to obtain extracts with higher phenolic content and antioxidant activity by studying methanol concentrations above 75% and extraction times longer than 9 min.

Table 4. Models for total extraction yield, total phenolic content and antioxidant activity (ABTS and DPPH) of matico leaves using UAE.

Response	Reparametrized Models	R^2 (%)
TEY	$Y = 12.31 + 0.3892X_1 + 0.3654X_2 - 0.00478{X_1}^2$	88.15
TPC	$Y = 3.937 - 0.03391X_1 - 0.215X_2 + 0.0006971{X_1}^2 + 0.005833X_1X_2$	97.05
ABTS	$Y = 73.22 - 1.335X_1 - 2.215X_2 + 0.01919{X_1}^2 + 0.04847X_1X_2$	96.45
DPPH	$Y = 5.88 - 0.0317X_1 + 1.08X_2 + 0.00803{X_1}^2$	93.76

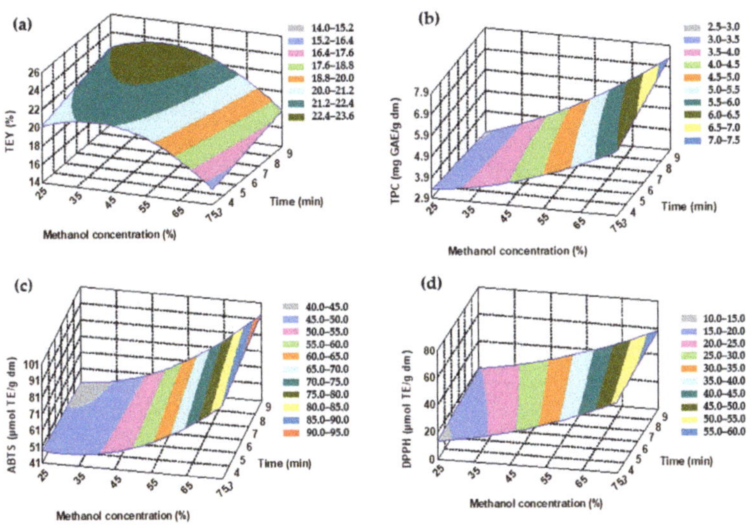

Figure 1. Response surface-matico leaf extracts: (**a**) total extraction yield (TEY), (**b**) total phenolic content (TPC), (**c**) antioxidant activity by ABTS, (**d**) antioxidant activity by DPPH. (The value of the third variable was set at the center point).

2.2.2. UAE of Chacruna (*Psychotria viridis* Ruiz & Pav.) Leaves

In the UAE of bioactive compounds from chacruna leaves, the impact of the variables was less noticeable than in the UAE of matico leaves; even for DPPH, the ANOVA showed that none of the terms were significant (Table S8). Significant and non-significant terms are presented in Tables S5–S8.

The significant effect of methanol concentration can be seen in experiments 1 and 2, where the values increased by varying the methanol concentration from 25% to 75% while keeping the other variables constant (from 17.3% to 21.2%, from 30.17 to 45.73 mg GAE/g dm, from 562.95 to 939.72 µmol TE/g dm, and from 365.51 to 547.55 µmol TE/g dm for TEY, TPC, ABTS, and DPPH, respectively).

Power increased the concentration of bioactive compounds in the UAE of chacruna leaves. Experiments 6 and 8 (Table 3) show the influence of this variable. When the power was increased from 90 W to 450 W, the results for TEY, TPC, ABTS, and DPPH also rose from 9.43% to 14.84%, 20.75 to 27.67 mg GAE/g dm, 422.41 to 534.68 µmol TE/g dm, and 276.08 to 418.65 µmol TE/g dm, respectively.

The content of bioactive compounds decreased with increasing time, as seen in experiments 2 and 4 (Table 3). Increasing the time from 3 to 9 min under the same conditions of methanol concentration and power significantly reduced the values achieved (21.2% to 17.86%, 45.73 to 31.83 mg GAE/g dm, 939.72 to 591.54 µmol TE/g dm, and 547.55 to 404.76 µmol TE/g dm for the values of TEY, TPC, ABTS, and DPPH, respectively).

The R^2 values of the models proposed for UAE of chacruna leaves were between 76.08% and 82.33% (Table 5), which were lower than those obtained for matico. The lack of fit for the TEY, TPC, and DPPH was not significant, indicating that the models adequately described the experimental data. However, the lack of fit for the ABTS response was significant (p-value = 0.0068), indicating that the proposed model did not adequately represent the data and, therefore, should not be used for prediction purposes. The optimization at the levels studied predicted a maximum value of 21.28% with 53.93% methanol, 3 min, and 321.06 W for TEY; 41.34 mg GAE/g dm with 70.82% methanol, 3 min, and 313.48 W for TPC; and 539.03 μmol TE/g dm under the conditions of 75% methanol, 3 min, and 432.66 W for DPPH.

Table 5. Models for total extraction yield, total phenolic content and antioxidant activity (ABTS and DPPH) of chacruna leaves using UAE.

Response	Non-Reparametrized Models	R^2 (%)
TEY	$Y = 10.53 + 0.3431X_1 - 1.769X_2 + 0.03691X_3 - 0.004099X_1^2 - 0.004767X_1X_2 + 0.0003528X_1X_3 + 0.1595X_2^2 - 0.0003519X_2X_3 - 0.00008547X_3^2$	76.08
TPC	$Y = 17.13 + 0.7411X_1 - 5.507X_2 + 0.07944X_3 - 0.005877X_1^2 - 0.0211X_1X_2 + 0.0004922X_1X_3 + 0.4713X_2^2 + 0.00009259X_2X_3 - 0.0001827X_3^2$	82.33
ABTS	$Y = 402.55 + 8.006X_1 - 79.07X_2 + 1.355X_3 - 0.02404X_1^2 - 0.7802X_1X_2 + 0.007339X_1X_3 + 8.38X_2^2 - 0.004366X_2X_3 - 0.002995X_3^2$	72.72
DPPH	$Y = 146.58 + 9.449X_1 - 35.43X_2 + 0.7412X_3 - 0.08246X_1^2 - 0.2538X_1X_2 + 0.008598X_1X_3 + 4.19X_2^2 - 0.04136X_2X_3 - 0.001457X_3^2$	80.04

When analyzing the response surfaces (Figure 2), small regions optimized at the studied levels for methanol concentration and power (Figure 2a–c) were observed, and their values coincided with those mentioned previously. There were no optimized regions in the graphs where time was involved, which follows a trend suggesting that studying times of less than 3 min could further improve the extraction process. For both matico and chacruna, the response surface graphs of antioxidants were very similar to those of TPC, which is explained by the strong correlation between TPC and the AA of the extracts, as explained in Section 2.3.

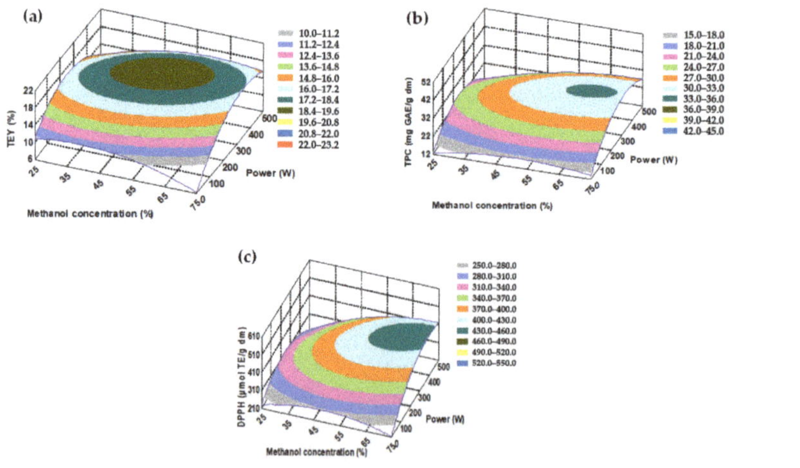

Figure 2. Response surface-chacruna leaf extracts: (**a**) total extraction yield (TEY), (**b**) total phenol content (TPC), (**c**) antioxidant activity by DPPH assay. (The value of the third variable was set at the center point).

2.2.3. Simultaneous Optimization of All Responses

To evaluate the extraction conditions that optimize all the response variables for matico and chacruna (using the models presented in Tables 4 and 5), an optimization study was conducted using the "Optimize Responses" option of Statgraphics Centurion software (18.1.13). The purpose was to estimate the extraction conditions that would be expected to yield high TEY, TPC, ABTS, and DPPH values. Table 6 shows the maximum values predicted by individual optimization and those that would theoretically be obtained under multivariable optimization.

Table 6. Predicted values of the responses obtained under optimal extraction conditions.

Response	Matico (*Piper aducum* L.) Leaves			Chacruna (*Psychotria viridis* Ruiz & Pav.) Leaves		
	Predicted Optimal Value	Predicted Value by Simultaneous Optimization *	A%	Predicted Optimal Value	Predicted Value by Simultaneous Optimization **	A%
TEY	23.53	18.33	77.92	21.28	20.83	97.84
TPC	7.32	7.16	97.76	41.34	40.86	98.84
ABTS †	93.85	91.27	97.25	-	-	-
DPPH	58.41	56.88	97.37	539.03	526.38	97.65

(A%) = Percentage achieved by simultaneous optimization over the expected optimal value; TEY (%); TPC (mg GAE/g dm); ABTS and DPPH (μmol TE/g dm). * 73.68% methanol, 9 min, 269.32 W. ** 64.84% methanol, 3 min, 344.44 W. † ABTS model for chacruna showed a significant lack of fit and was excluded from optimization.

For the simultaneous optimization of the chacruna data, the ABTS response was excluded because, as previously noted, its model showed a significant lack of fit, making it unsuitable for prediction. Therefore, the optimization results (Table 6) represent theoretical values corresponding to the predicted optimal conditions, which could be experimentally validated to confirm the reliability of the models.

2.3. *Correlation Between TPC and AA of Matico and Chacruna Leaf Extracts*

Figure 3 shows that the Pearson correlation coefficients (r) between AA (measured by ABTS and DPPH) and TPC ranged from 0.9321 to 0.9700. Therefore, we can affirm that phenolic compounds are primarily responsible for the antioxidant activity in the extracts of matico and chacruna leaves.

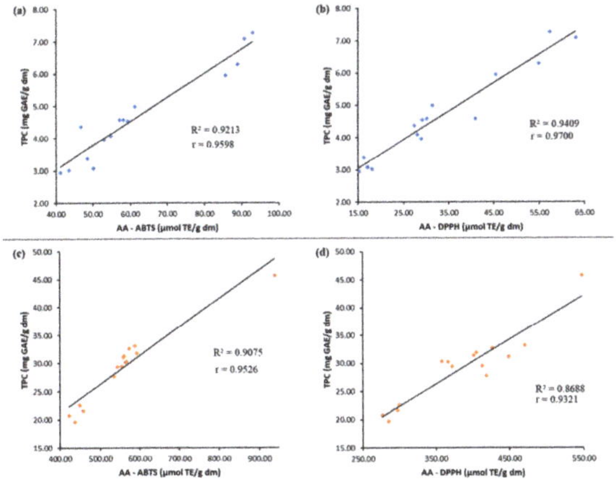

Figure 3. Correlation between the antioxidant activity (AA) and the total phenolic content (TPC) of the extracts obtained by UAE under different conditions. Matico leaf extracts (**a**,**b**). Chacruna leaf extracts (**c**,**d**).

3. Discussion

Since no reports had been found that allowed for comparison of the characterization of matico and chacruna leaves, a search was conducted on the proximal composition of other species belonging to the genera *Piper* and *Psychotria*. The dried leaves of *Piper guineense* showed 18.19% crude fiber, 2.45% fat, 11.55% moisture, 11.32% protein, 14.25% ash, and 42.23% carbohydrates [23]. Meanwhile, the dried leaves of *Psychotria* sp. contained 3.40% crude fiber, 8.55% fat, 12.87% moisture, 27.32% protein, 7.06% ash, and 40.80% carbohydrates [24]. Some of the mentioned data were relatively similar; however, the variability could be explained by the influence of factors such as species, soil, climatic conditions, altitude, and collection time [25]. These factors caused significant qualitative and quantitative changes in the phytochemical profile of these plants [6,21].

Most studies on matico and chacruna refer to phytochemical analyses. Matico leaf extracts have been found to contain a high concentration of flavonoids, followed by a moderate concentration of phenols, alkaloids, and terpenoids [26]. Certain phenols and flavonoids, including gallic acid, catechin, chlorogenic acid, caffeic acid, ellagic acid, rutin, quercetin, luteolin, and apigenin, have been detected in chacruna extracts [27]. Various factors influenced the extraction of active compounds, such as the solvent, solute:solvent ratio, time, temperature, ultrasonic power, and particle size. The influence of these parameters varied depending on the plant and the responses analyzed [28].

When working with response surface modeling, a high R^2 and a non-significant lack of fit ($p > 0.05$) are indicators that the quadratic model is reliable and that the variation between samples is due only to the factors selected by the model and pure error [29].

In the UAE of matico (*Piper aduncum* L.) leaves, the lack of significance of the power variable is likely due to the fact that the particles were very small (powdered) and did not hinder mass transfer [25]. In addition, the methanol/water mixture destroys cell membranes, allowing the release and stabilization of certain phenolic subgroups. Likewise, methanol mitigates the degradation of phenols by inhibiting the activity of polyphenol oxidases, which are abundant in plants [30]. The positive effect of time was likely due to greater contact between the solvent and the plant material, which facilitated the diffusion of the tissue components into the solvent [31]. This may be related to an increase in the formation of microbubbles, which intensify cellular damage by facilitating the release of phytochemicals [32].

Only for TPC, ANOVA showed a significant interaction between methanol concentration and extraction time (Table S2). According to the fitted model (Table 4), this interaction was positive, indicating that higher methanol concentrations combined with longer extraction times resulted in greater recovery of phenolic compounds. This behavior may be attributed to the enhanced solubility and diffusion of phenolics when sufficient extraction time is provided.

Other authors have reported results for TPC and AA of matico leaf extracts. However, because the results were expressed in various ways, comparing them with those obtained in this study is difficult. Arroyo et al. [33] reported a TPC of around 16 mg GAE/mL for the methanolic fraction of an extract; Yarleque-Chocas et al. [1] reported a value of 21.12 mg GAE/g sample using 96% ethanol through maceration for seven days; Luna-Fox et al. [34] reported values of 2.24 g GAE/100 g dm in an aqueous extract using low-power ultrasound, and antioxidant activity measured by the ABTS assay of 86.4 mg TE/100 g dm, respectively.

It was not convenient to reparameterize the polynomial models for the chacruna leaf data. Sometimes, when only the significant terms are considered, there can be a high error between the experimental and predicted data; therefore, it is more appropriate to use complete polynomial models [20].

In the UAE of chacruna (*Psychotria viridis* Ruiz & Pav.) leaves, the negative impact of the quadratic effect of power on TEY, TPC, and DPPH indicates that extraction improves as power increases up to a certain point and then begins to decrease. This could be attributed to the possible degradation of compounds caused by the ultrasonic power supplied, which raises the system's temperature due to the magnitude of the cavitation generated [35].

The negative impact of time on TEY, TPC, and DPPH may result from the thermal degradation of the compounds present in chacruna leaves [36]. This degradation is caused by the combination of high power and prolonged times, which triggers high temperatures and pressures [31]. Prolonged exposure to ultrasound can cause adverse effects, such as reduced levels of phenolic compounds due to forming free radicals [36,37]. This leads to chemical decomposition, negatively affecting free radical scavenging activity [38]. It has also been reported that prolonged extraction periods reduce phenol levels due to oxidation [36].

The lower R^2 values in the models obtained for chacruna compared with those achieved for matico could be due to the variability of the compounds, including phenolics since these molecules have various polarities and sizes, ranging from simple phenols to tannins [19].

To our knowledge, the content of phenolic compounds and the antioxidant activity of chacruna leaves have not yet been investigated. For comparative purposes, a study was found on extracts from *Psychotria nilgiriensis* leaves, where the authors reported 66.11 mg GAE/g extract with methanol as a solvent; the antioxidant activity was 833.3 μg/mL by DPPH and 1277.43 μM Trolox equivalent/g extract by ABTS [39].

In general, the highest TEY values for matico and chacruna were achieved with 50% methanol, whereas the phenolic compounds were extracted in greater amounts with 75% methanol concentrations (Tables 2 and 3). Mixtures of solvents with water increase diffusivity and improve the extraction rate [40] A higher water content also enhances the extraction of certain contaminant solutes, affecting the concentration of phenolic compounds [41]. In other words, decreasing the concentration of methanol will result in extracts with a high yield (due to the presence of contaminating materials) but with lower antioxidant activity [42].

Both species studied show variability in their content of bioactive compounds compared to other studies. Factors such as species, soil, climatic conditions, altitude, and collection time cause these differences [25]. It is also difficult to compare with other reports because the results are not expressed uniformly.

The solvent concentration is one of the most significant variables in the UAE [42]. This was reflected in our results, where the methanol concentration was the variable that most affected the extraction of bioactive compounds from matico and chacruna leaves. The mixture of methanol and water decreases viscosity, facilitating cavitation [38]. The maximized conditions (Table 6) coincide with those reported in other studies. For example, Yerena-Prieto et al. [37] observed that the methanol concentration is the most influential variable in extracting bioactives from moringa leaves.

Several studies have found a proportional relationship between the increase in solvent concentration up to a certain point (solvent/water mixtures) and the content of phenolic compounds [17,19,20]. However, it is necessary to carry out studies to confirm which solvent is the most suitable since the solubility of TPC cannot be explained solely by its polarity, other factors also influence it, such as stereochemistry, intermolecular forces with the solvent [43], and the nature of the plant itself [25].

The bioactive compounds of the matico were not affected by the ultrasonic power or the increase in time, indicating good thermal stability. At the same time, these variables degraded the phytochemicals of the chacruna, indicating that they would be more sensitive to temperature. This behavior can be explained by the thermostability of phenolic compounds, which depends on their chemical nature [44].

The ABTS values were higher than those of DPPH (Tables 2 and 3), possibly due to the presence of various phenolic compounds with different antioxidant capacities and the varying mechanisms of action of the antioxidant assays [45].

Consistent with our findings several studies have pointed out an important dependence between the content of phenolic compounds and antioxidant activity [20,42]. It is essential to know if the phenolic compounds in the extracts exhibit antioxidant activity, because this characteristic is linked to health benefits and is highly sought after in various industries [20].

4. Materials and Methods

4.1. Material

Fresh leaves of *Piper aduncum* L. (matico) and *Psychotria viridis* Ruiz & Pav. (chacruna) were collected from an experimental center located in the cities of Rioja and Tarapoto, respectively (San Martín, Peru), in September 2022. In both cases, the methodology proposed by Farahmandfar et al. [40] was followed. The leaves were separated, cleaned, and dried under shade at room temperature for one week. The samples were then ground using a blade mill (OSTER) to homogenize the material. The milled leaves were placed in airtight glass jars and refrigerated at 2 °C until further analysis.

ABTS (2,2′-azino-bis-(3-ethylbenzothiazoline)-6-sulfonic acid), DPPH (1,1-diphenyl-2-picryl- hydrazyl), potassium persulfate (CAS 7727-21-1), anhydrous gallic acid (CAS 149-91-7), and Trolox were purchased from Merck KGaA (Darmstadt, Germany). Sodium carbonate (CAS 144-55-8) and Folin–Ciocalteu reagent were purchased from Tawa (Lima, Peru). Methanol (CAS 67-56-1, purity 99.8%) was purchased from Grupo Química (Lima, Peru). Distillation equipment was used for the water (2008, GFL, Burgwedel, Germany).

4.2. Characterization of Raw Material

The dried and crushed leaves were characterized in terms of crude fiber using the Peruvian Technical Standard (NTP 205.003:1980) [46], and fat, moisture, protein, and ash were determined by methods 930.09, 930.04, 978.04 (A), and 930.05, respectively, according to AOAC (2019) [47]. Carbohydrates were estimated by difference, subtracting the sum of moisture, protein, fat, and ash from 100 g of sample, and total energy was calculated using the Atwater factors, assigning 4 kcal/g for protein, 4 kcal/g for carbohydrates, and 9 kcal/g for fat [48].

4.3. Ultrasound-Assisted Extraction (UAE)

The extraction was performed using an ultrasonic probe equipment (900 W power, 12.5 mm Ø, frequency 20 kHz, Biobase, Jinan, China). 3 g of sample were mixed with 50 mL of the extracting solvent (methanol/water in different concentrations). After UAE, the extract was centrifuged (EPPENDORF, 5702, Hamburg, Germany) at 3500 rpm for 5 min. The supernatant was collected and stored in amber bottles at 2 °C until further characterization.

4.4. Box–Behnken Experimental Design

The UAE of bioactive compounds from matico and chacruna leaves was carried out according to a Box–Behnken experimental design. The experimental conditions were generated with Statgraphics Centurion software, giving 15 experiments (3 center points and 12 combinations of independent variables, Tables 2 and 3). The independent parameters were the methanol concentration (X_1), extraction time (X_2) and power (X_3). The experimental data was fitted with a second-order polynomial equation (Equation (1)) to maximize the response variables (TEY, TPC, ABTS, and DPPH).

$$Y = \beta_0 + \sum \beta_i x_i + \sum \beta_{ii} x_{ii}^2 + \sum \beta_{ij} x_i x_j + \varepsilon \quad (1)$$

where Y is the response variable, β_0 is an independent parameter, β_i is the linear effect of the input factor x_i, β_{ii} is the quadratic effect of the input factor x_i, and β_{ij} represents the linear interaction effect between the input factor input x_i and x_j, ε symbolizes the error. The suitability of the models was evaluated through the multiple regression coefficients (R^2) and lack of fit [17].

New models were generated for the UAE of matico leaves by eliminating coefficients that were not significant ($p > 0.05$). Meanwhile, for chacruna leaves, the models were analyzed including all the terms, because when attempting to reparameterize, very low R^2 and some significant lack of fit were obtained ($p < 0.05$).

4.5. Extract Characterization

4.5.1. Total Extraction Yield (TEY)

To determine the TEY, the methodology proposed by Vélez-Erazo et al. [49] was followed. 3 mL of extract were placed in a Petri dish and dried in a conventional oven (Venticell 111-ECO line, BMT medical, Kassel, Germany) at 105° C, evaporating the solvent until a constant weight was obtained. The results were calculated with Equation (2).

$$TEY\ (\%) = \frac{W_1 \times 100}{W_2} \quad (2)$$

where W_1 = weight of the dry extract, and W_2 = initial weight of the sample. The weights were determined on an analytical balance (Pioneer TM PX224, OHAUS, Mexico City, Mexico).

4.5.2. Total Phenolic Content (TPC)

The total phenolic content was determined using the Folin–Ciocalteu spectrophotometric method described by Singleton et al. [50], with some modifications. In brief, 200 µL of extract, 200 µL of diluted Folin–Ciocalteu reagent (1:2), 400 µL of sodium carbonate (10% *w*/*v*), and 3200 µL of distilled water were mixed, stirred for 5 s, and incubated for 30 min at 25 °C in the dark. The absorbance was measured at 765 nm in a spectrophotometer (S-220 UV/VIS, BOECO, Hamburg, Germany). The calibration curve was prepared with a standard solution of gallic acid (0–0.12 mg/mL). Equation (3) was used to calculate the results [51].

$$Tpc = \frac{c \times V}{m} \quad (3)$$

where c = concentration of the sample in the calibration curve, V = volume of the solvent used, and m = sample weight. The results were expressed as mg of gallic acid equivalents (GAE) per gram of dry matter (dm).

4.5.3. Antioxidant Activity-ABTS (Ammonium 2,2′-Azino-bis-(3-ethyl benzothiazolin-6-sulfonate))

The antioxidant activity by ABTS assay was determined through the spectrophotometric method described by Re et al. [52], with certain modifications. 5 mL of ABTS aqueous solution (7 mM) were reacted with 5 mL of potassium persulfate solution (2.45 mM) and kept in the dark for 16 h. Then, the prepared solution was diluted in ethanol in a 1:90 ratio and adjusted to an absorbance of 0.700 ± 0.02 at 734 nm. 20 µL of extract and 2 mL of $ABTS^+$ solution were mixed, incubated for 6 min, and then the absorbance was measured at 734 nm. The calibration curve was constructed with a standard Trolox solution (500–2000 μM). The antioxidant activity of the extracts was expressed as μmol Trolox equivalent (μmol TE)/g dm.

4.5.4. Antioxidant Activity-DPPH (1,1-Diphenyl-2-picryl-hydrazil)

Antioxidant activity by DPPH assay was evaluated according to Brand-Williams et al. [53], with some adaptations. The calibration curve was constructed with a standard Trolox solution (20–200 μM). First, the DPPH solution was calibrated to an absorbance of 0.800 ± 0.02 at 515 nm. Then, 500 μL of extract, 3 mL of ethanol, and 300 μL of DPPH solution were added, shaken for 5 s and incubated in the dark for 45 min. The percentage of radical scavenging activity (*RSA*) was calculated using Equation (4).

$$(\%)RSA = \left(\frac{Abs_{control} - Abs_{sample}}{Abs_{control}}\right) \times 100 \quad (4)$$

where $Abs_{control}$ = initial absorbance of the methanolic DPPH solution $Abs_{muestra}$ = absorbance of the reaction mixture (DPPH + sample). The AA of the extracts was expressed as μmol Trolox equivalent (μmol TE)/g dm.

4.6. Statistical Analysis

Three experimental repetitions were carried out for each treatment, expressed as average ± standard deviation. Statgraphics Centurion software (trial version 18.1.13, Stat Point Technologies, Inc., Richmond, VA, USA) was used for statistical analysis of the results by RSM. The independent variables' linear, quadratic, and interaction effects on the responses were established with the analysis of variance (ANOVA) with a significance level of 5%. The adequacy and precision of the models were evaluated using the multiple regression coefficient (R^2) and lack of fit. To obtain the values of methanol concentration (X_1), time (X_2), and power (X_3) that allow for maximizing the results, the "Optimize Response" software option was used. This was done for each response variable separately, followed by simultaneous optimization. Pearson's correlation coefficient was used to determine the linear correlation between TPC and AA with Microsoft Office Excel 2019 software.

5. Conclusions

In the extracts of matico leaves, methanol concentration was the most significant variable, followed by time, while power had no effect. Likewise, the reparametrized quadratic models adequately adjusted the experimental results and could be used to make predictions. The response surface graphs suggest that the extraction process of TPC and antioxidants could be improved with methanol concentrations greater than 75% and extraction times longer than 9 min. The multiresponse analysis at the levels studied, indicated that under the conditions of 73.68% methanol, 9 min, and 269.32 W, up to 18.33%, 7.16 mg GAE/g dm, 91.27 μmol TE/g dm and 56.88 μmol TE/g dm of TEY, TPC, ABTS, and DPPH would be obtained, respectively.

Power, methanol concentration, and time influenced the UAE process of chacruna leaves. The quadratic models were not reparameterized, resulting in lower R^2 values than those for matico. However, the lack of fit was not significant, demonstrating the models' reliability (except for the polynomial for the ABTS response), and the ANOVA of DPPH did not show significant terms (p-value < 0.05); therefore, other variables, such as the solute/solvent ratio, applying pre-treatments, other solvents, and even different methodologies or experimental designs, should be studied. The response surface graphs also show the trend that extraction could be improved with times of less than 3 min. The optimization conditions in the multi-response analysis at the levels studied were 64.84%

methanol, 3 min, 344.44 W, with the model predicting maximum values of TEY, TPC and DPPH (20.83%, 40.86 mg GAE/g dm, and 526.38 μmol TE/g dm, respectively), excluding the ABTS model due to its significant lack of fit.

A strong correlation was observed between the antioxidant activity of the matico and chacruna extracts and their total phenolic content. Ultrasound-assisted extraction can be used to obtain natural antioxidants from matico and chacruna leaves, which could help develop new products in various fields of industry. In future studies, the methodological framework proposed in this work could be applied using environmentally friendly solvents, such as ethanol–water systems, to enhance the sustainability and industrial applicability of the optimized UAE process. The data obtained could serve as a basis to support the medicinal use of these plants; however, more research is required to isolate and characterize the phytochemicals responsible for their medicinal properties.

Supplementary Materials: The following supporting information can be downloaded at: https://www.mdpi.com/article/10.3390/molecules30224395/s1, Table S1. Analysis of Variance for TEY (%) of Matico. Table S2. Analysis of variance for TPC (mg GAE/g dm) of Matico. Table S3. Analysis of variance for ABTS (μmol TE/g dm) of Matico. Table S4. Analysis of variance for DPPH (μmol TE/g dm) of Matico. Table S5. Analysis of variance for TEY (%) of Chacruna. Table S6. Analysis of variance for TPC (mg GAE/g dm) of Chacruna. Table S7. Analysis of variance for ABTS (μmol TE/g dm) of Chacruna. Table S8. Analysis of variance for DPPH (μmol TE/g dm) of Chacruna.

Author Contributions: R.R.-S.: conceptualization, formal analysis, investigation. R.P.-V.: visualization, Writing—original draft. J.L.P.-R.: project administration, Writing—review and editing, visualization, supervision, Conceptualization. M.F.C.-J.: funding acquisition, methodology. P.V.-R.: project administration, resources. Á.C.-G.: data analysis and interpretation. K.S.-D.: supervision—review and editing. K.R.-D.Á.: identification and characterization of the raw material. All authors have read and agreed to the published version of the manuscript.

Funding: The authors thank the Universidad Nacional de San Martín-PERU for the financial support according to Resolution N° 623-2022-UNSM/CU-R and Resolution N° 611-2022-UNSM/CU-R.

Institutional Review Board Statement: Not applicable.

Informed Consent Statement: Not applicable.

Data Availability Statement: The original contributions presented in this study are included in the article/Supplementary Materials. Further inquiries can be directed to the corresponding author.

Conflicts of Interest: None of the authors has any conflict of interest in this research.

References

1. Yarleque-Chocas, M.; Dorregaray-Llerena, F.; Yarleque-Chocas, A.; Gonzales-Chavesta, C. Actividad Antiinflamatoria in Vitro de *Plantago major* L. y *Piper aduncum* L. Sobre la Fosfolipasa A2 del Veneno de la Serpiente *Lachesis muta muta*. *Rev. Peru. Med. Exp. Salud Publica* **2023**, *40*, 325–332. [CrossRef] [PubMed]
2. Efdi, M.; Okselni, T.; Itam, A.; Arifin, B.; Novela, M.; Hidayat, T. Fadli Essential Oil Extraction of *Piper betle*, *Piper ramipilum*, and *Piper aduncum* and Their Antibacterial Activity Against Food Borne Pathogens. *J. Essent. Oil Bear. Plants* **2023**, *26*, 446–458. [CrossRef]
3. Ahmad Khan, M.S.; Ahmad, I. Herbal Medicine: Current Trends and Future Prospects. In *New Look to Phytomedicine: Advancements in Herbal Products as Novel Drug Leads*; Ahmad Khan, M.S., Ahmad, I., Chattopadhyay, D., Eds.; Academic Press: Cambridge, MA, USA, 2019; pp. 3–13, ISBN 9780128146200.
4. Rivero-Montejo, S.d.J.; Vargas-Hernandez, M.; Torres-Pacheco, I. Nanoparticles as Novel Elicitors to Improve Bioactive Compounds in Plants. *Agriculture* **2021**, *11*, 134. [CrossRef]
5. Gali, L.; Bedjou, F.; Velikov, K.P.; Ferrari, G.; Donsì, F. High-Pressure Homogenization-Assisted Extraction of Bioactive Compounds from *Ruta chalepensis*. *J. Food Meas. Charact.* **2020**, *14*, 2800–2809. [CrossRef]

6. Herrera-Calderon, O.; Chacaltana-Ramos, L.; Yuli-Posadas, R.Á.; Pari-Olarte, B.; Enciso-Roca, E.; Tinco-Jayo, J.A.; Rojas-Armas, J.P.; Felix-Veliz, L.M.V.; Franco-Quino, C. Antioxidant and Cytoprotective Effect of *Piper aduncum* L. against Sodium Fluoride (NaF)-Induced Toxicity in Albino Mice. *Toxics* **2019**, *7*, 28. [CrossRef]
7. Saldaña-Chafloque, C.F.; Mostacero-León, J.; De la Cruz-Castillo, A.J. Plantas Medicinales Empleadas en el Tratamiento de Enfermedades del Sistema Respiratorio Por la Comunidad Andina de Pampas, Tayacaja, Huancavelica. *TAYACAJA* **2022**, *5*, 25–32. [CrossRef]
8. Gaujac, A.; Navickiene, S.; Collins, M.I.; Brandt, S.D.; de Andrade, J.B. Analytical Techniques for the Determination of Tryptamines and β-Carbolines in Plant Matrices and in Psychoactive Beverages Consumed During Religious Ceremonies and Neo-Shamanic Urban Practices. *Drug Test. Anal.* **2012**, *4*, 636–648. [CrossRef]
9. Moraes, T.F.S.; Ferraz, A.C.; da Cruz Nizer, W.S.; Tótola, A.H.; Soares, D.B.S.; Duarte, L.P.; Vieira-Filho, S.A.; Magalhães, C.L.B.; de Magalhães, J.C. A Methanol Extract and N,N-Dimethyltryptamine from *Psychotria viridis* Ruiz & Pav. Inhibit *Zika virus* Infection In Vitro. *Arch. Virol.* **2021**, *166*, 3275–3287. [CrossRef]
10. Kaasik, H.; Souza, R.C.Z.; Zandonadi, F.S.; Tófoli, L.F.; Sussulini, A. Chemical Composition of Traditional and Analog Ayahuasca. *J. Psychoact. Drugs* **2021**, *53*, 65–75. [CrossRef]
11. Luca, S.V.; Macovei, I.; Bujor, A.; Miron, A.; Skalicka-Woźniak, K.; Aprotosoaie, A.C.; Trifan, A. Bioactivity of Dietary Polyphenols: The Role of Metabolites. *Crit. Rev. Food Sci. Nutr.* **2019**, *60*, 626–659. [CrossRef]
12. Manousi, N.; Sarakatsianos, I.; Samanidou, V. Extraction Techniques of Phenolic Compounds and Other Bioactive Compounds from Medicinal and Aromatic Plants. In *Engineering Tools in the Beverage Industry*; Grumezescu, A.M., Holban, A.M., Eds.; Woodhead Publishing: Cambridge, UK, 2019; pp. 283–314, ISBN 9780128152584.
13. Ivanović, M.; Razboršek, M.I.; Kolar, M. Innovative Extraction Techniques Using Deep Eutectic Solvents and Analytical Methods for the Isolation and Characterization of Natural Bioactive Compounds from Plant Material. *Plants* **2020**, *9*, 1428. [CrossRef]
14. Khoddami, A.; Wilkes, M.A.; Roberts, T.H. Techniques for Analysis of Plant Phenolic Compounds. *Molecules* **2013**, *18*, 2328–2375. [CrossRef] [PubMed]
15. Rutkowska, M.; Namieśnik, J.; Konieczka, P. Ultrasound-Assisted Extraction. In *The Application of Green Solvents in Separation Processes*; Pena-Pereira, F., Tobiszewski, M., Eds.; Elsevier Inc.: Amsterdam, The Netherlands, 2017; pp. 301–324. ISBN 9780128054437.
16. Weremfo, A.; Abassah-Oppong, S.; Adulley, F.; Dabie, K.; Seidu-Larry, S. Response Surface Methodology as a Tool to Optimize the Extraction of Bioactive Compounds from Plant Sources. *J. Sci. Food Agric.* **2022**, *103*, 26–36. [CrossRef] [PubMed]
17. Ganesan, V.; Gurumani, V.; Kunjiappan, S.; Panneerselvam, T.; Somasundaram, B.; Kannan, S.; Chowdhury, A.; Saravanan, G.; Bhattacharjee, C. Optimization and Analysis of Microwave-Assisted Extraction of Bioactive Compounds from *Mimosa pudica* L. Using RSM & ANFIS Modeling. *J. Food Meas. Charact.* **2018**, *12*, 228–242. [CrossRef]
18. Nuapia, Y.; Cukrowska, E.; Tutu, H.; Chimuka, L. Statistical Comparison of Two Modeling Methods on Pressurized Hot Water Extraction of Vitamin C and Phenolic Compounds from *Moringa oleifera* Leaves. *South Afr. J. Bot.* **2020**, *129*, 9–16. [CrossRef]
19. Fernández-Barbero, G.; Pinedo, C.; Espada-Bellido, E.; Ferreiro-González, M.; Carrera, C.; Palma, M.; García-Barroso, C. Optimization of Ultrasound-Assisted Extraction of Bioactive Compounds from Jabuticaba (*Myrciaria cauliflora*) Fruit through a Box-Behnken Experimental Design. *Food Sci. Technol.* **2019**, *39*, 1018–1029. [CrossRef]
20. Tranquilino-Rodríguez, E.; Martínez-Flores, H.E.; Rodiles-López, J.O.; Figueroa-Cárdenas, J.D.D.; Pérez-Sánchez, R.E. Optimization in the Extraction of Polyphenolic Compounds and Antioxidant Activity from Opuntia Ficus-Indica Using Response Surface Methodology. *J. Food Process Preserv.* **2020**, *44*, e14485. [CrossRef]
21. Morais, V.P.; Cabral, F.V.; Fernandes, C.C.; Miranda, M.L.D. Brief Review on *Piper aduncum* L., Its Bioactive Metabolites and Its Potential to Develop Bioproducts. *Braz. Arch. Biol. Technol.* **2023**, *66*, e23220314. [CrossRef]
22. Zhang, G.; He, L.; Hu, M. Optimized Ultrasonic-Assisted Extraction of Flavonoids from *Prunella vulgaris* L. and Evaluation of Antioxidant Activities In Vitro. *Innov. Food Sci. Emerg. Technol.* **2011**, *12*, 18–25. [CrossRef]
23. Ojinnaka, M.C.; Odimegwu, E.N.; Chidiebere, F.E. Comparative Study on the Nutrient and Antinutrient Composition of the Seeds and Leaves of Uziza (*Piper guineense*). *IOSR J. Environ. Sci.* **2016**, *10*, 42–48. [CrossRef]
24. Otitoju, G.; Ene-Obong, H.; Otitoju, O. Macro and Micro Nutrient Composition of Some Indigenous Green Leafy Vegetables in South-East Zone Nigeria. *J. Food Process Technol.* **2014**, *5*, 389. [CrossRef]
25. Bouafia, M.; Colak, N.; Ayaz, F.A.; Benarfa, A.; Harrat, M.; Gourine, N.; Yousfi, M. The Optimization of Ultrasonic-Assisted Extraction of *Centaurea* Sp. Antioxidative Phenolic Compounds Using Response Surface Methodology. *J. Appl. Res. Med. Aromat. Plants* **2021**, *25*, 100330. [CrossRef]
26. Rachmaini, F.; Abdillah, R.; Oktavia, S.; Wahyuni, F.S. Cholesterol Lowering Acitivity and Vasorelaxant Effect of Ethanol Extract *Piper aduncum* Leaves in Hypercholesterolemic Wistar Kyoto Rats. *S. Afr. J. Bot.* **2024**, *164*, 366–373. [CrossRef]
27. Andrade, A.O.; da Silva, M.A.P.; de Oliveira, A.H.; dos Santos, M.A.F.; Vandesmet, L.C.S.; Generino, M.E.M.; Coelho, H.K.R.C.; Landim, H.S.; Mendonça, A.C.A.M.; da Costa, N.C. Allelopathic Effects of *Psychotria viridis* Ruiz & Pavon on the Germination and Initial Growth of *Lactuca sativa* L. *J. Agric. Sci.* **2017**, *9*, 204–217. [CrossRef]

28. Chen, X.; Ding, J.; Ji, D.; He, S.; Ma, H. Optimization of Ultrasonic-Assisted Extraction Conditions for Bioactive Components from Coffee Leaves Using the Taguchi Design and Response Surface Methodology. *J. Food Sci.* **2020**, *85*, 1742–1751. [CrossRef]
29. Puértolas, E.; Saldaña, G.; Álvarez, I.; Raso, J. Experimental Design Approach for the Evaluation of Anthocyanin Content of Rosé Wines Obtained by Pulsed Electric Fields. Influence of Temperature and Time of Maceration. *Food Chem.* **2011**, *126*, 1482–1487. [CrossRef]
30. Vajić, U.J.; Grujić-Milanović, J.; Živković, J.; Šavikin, K.; Godevac, D.; Miloradović, Z.; Bugarski, B.; Mihailović-Stanojević, N. Optimization of Extraction of Stinging Nettle Leaf Phenolic Compounds Using Response Surface Methodology. *Ind. Crops Prod.* **2015**, *74*, 912–917. [CrossRef]
31. Yancheshmeh, B.S.; Panahi, Y.; Allahdad, Z.; Abdolshahi, A.; Zamani, Z. Optimization of Ultrasound-Assisted Extraction of Bioactive Compounds from *Achillea kellalensis* Using Response Surface Methodology. *J. Appl. Res. Med. Aromat. Plants* **2022**, *28*, 100355. [CrossRef]
32. Kazemi, M.; Karim, R.; Mirhosseini, H.; Abdul Hamid, A. Optimization of Pulsed Ultrasound-Assisted Technique for Extraction of Phenolics from Pomegranate Peel of Malas Variety: Punicalagin and Hydroxybenzoic Acids. *Food Chem.* **2016**, *206*, 156–166. [CrossRef]
33. Arroyo, J.; Bonilla, P.; Tomás, G.; Huamán, J. Estudio Fitoquímico del Extracto Etanólico y de las Fracciones de las Hojas de Piper Aduncum "Matico". *Rev. Peru. Química Ing. Química* **2011**, *14*, 62–67.
34. Luna-Fox, S.B.; Rodríguez-Almeida, N.; Dahua-Gualinga, R.D.; Peñafiel-Bonilla, N.J.; Radice, M. Influencia de las Condiciones de Extracción Sobre Los Compuestos Polifenólicos Totales del *Piper aduncum* L. *Cienc. Lat. Rev. Científica Multidiscip.* **2023**, *7*, 3872–3891. [CrossRef]
35. Nutter, J.; Fernandez, M.V.; Jagus, R.J.; Agüero, M.V. Development of an Aqueous Ultrasound-Assisted Extraction Process of Bioactive Compounds from Beet Leaves: A Proposal for Reducing Losses and Increasing Biomass Utilization. *J. Sci. Food Agric.* **2021**, *101*, 1989–1997. [CrossRef] [PubMed]
36. Benarfa, A.; Gourine, N.; Hachani, S.; Harrat, M.; Yousfi, M. Optimization of Ultrasound-Assisted Extraction of Antioxidative Phenolic Compounds from *Deverra scoparia* Coss. & Durieu (Flowers) Using Response Surface Methodology. *J. Food Process Preserv.* **2020**, *44*, e14514. [CrossRef]
37. Yerena-Prieto, B.J.; Gonzalez-Gonzalez, M.; Vázquez-Espinosa, M.; González-De-peredo, A.V.; García-Alvarado, M.Á.; Palma, M.; Rodríguez-Jimenes, G.D.C.; Barbero, G.F. Optimization of an Ultrasound-Assisted Extraction Method Applied to the Extraction of Flavonoids from Moringa Leaves (*Moringa oleífera* Lam.). *Agronomy* **2022**, *12*, 261. [CrossRef]
38. Kumar, A.; Srinivasa Rao, P. Optimization of Pulsed-Mode Ultrasound Assisted Extraction of Bioactive Compounds from Pomegranate Peel Using Response Surface Methodology. *J. Food Meas. Charact.* **2020**, *14*, 3493–3507. [CrossRef]
39. Iniyavan, M.; Sangeetha, D.; Saravanan, S.; Parimelazhagan, T. Evaluation of Antioxidant and Pharmacological Properties of *Psychotria nilgiriensis* Deb & Gang. *Food Sci. Biotechnol.* **2012**, *21*, 1421–1431. [CrossRef]
40. Farahmandfar, R.; Esmaeilzadeh Kenari, R.; Asnaashari, M.; Shahrampour, D.; Bakhshandeh, T. Bioactive Compounds, Antioxidant and Antimicrobial Activities of Arum Maculatum Leaves Extracts as Affected by Various Solvents and Extraction Methods. *Food Sci. Nutr.* **2019**, *7*, 465–475. [CrossRef]
41. Spigno, G.; Tramelli, L.; De Faveri, D.M. Effects of Extraction Time, Temperature and Solvent on Concentration and Antioxidant Activity of Grape Marc Phenolics. *J. Food Eng.* **2007**, *81*, 200–208. [CrossRef]
42. Almusallam, I.A.; Mohamed Ahmed, I.A.; Babiker, E.E.; Al Juhaimi, F.Y.; Fadimu, G.J.; Osman, M.A.; Al Maiman, S.A.; Ghafoor, K.; Alqah, H.A.S. Optimization of Ultrasound-Assisted Extraction of Bioactive Properties from Date Palm (*Phoenix dactylifera* L.) Spikelets Using Response Surface Methodology. *LWT* **2021**, *140*, 110816. [CrossRef]
43. Galanakis, C.M.; Goulas, V.; Tsakona, S.; Manganaris, G.A.; Gekas, V. A Knowledge Base for the Recovery of Natural Phenols with Different Solvents. *Int. J. Food Prop.* **2013**, *16*, 382–396. [CrossRef]
44. Setyaningsih, W.; Saputro, I.E.; Palma, M.; Barroso, C.G. Stability of 40 Phenolic Compounds during Ultrasound-Assisted Extractions (UAE). *AIP Conf. Proc.* **2016**, *1755*, 080009. [CrossRef]
45. Pandey, A.; Belwal, T.; Sekar, K.C.; Bhatt, I.D.; Rawal, R.S. Optimization of Ultrasonic-Assisted Extraction (UAE) of Phenolics and Antioxidant Compounds from Rhizomes of Rheum Moorcroftianum Using Response Surface Methodology (RSM). *Ind. Crops Prod.* **2018**, *119*, 218–225. [CrossRef]
46. Instituto Nacional de Calidad (INACAL). *Norma Técnica Peruana NTP 205.003:1980 (Revisada en 2011)*; Cereales y menestras. Determinación de la fibra cruda; Diario Oficial El Peruano: Lima, Peru, 2011.
47. AOAC. *Official Methods of Analysis. Chemistry*, 21st ed.; Association of Analytical Communities, Ed.; Association of Official Analytical Chemists: Washington, DC, USA, 2019.
48. Collazos, C. *La Composición de Alimentos de Mayor Consumo en el Perú*, 6th ed.; Ministerio de Salud, Instituto Nacional de Nutricion: Lima, Peru, 1993.

49. Vélez-Erazo, E.M.; Pasquel-Reátegui, J.L.; Dorronsoro-Guerrero, O.H.; Martínez-Correa, H.A. Phenolics and Carotenoids Recovery from Agroindustrial Mango Waste Using Microwave-Assisted Extraction: Extraction and Modeling. *J. Food Process Eng.* **2021**, *44*, e13774. [CrossRef]
50. Singleton, V.L.; Orthofer, R.; Lamuela-Raventós, R.M. Analysis of Total Phenols and Other Oxidation Substrates and Antioxidants by Means of Folin-Ciocalteu Reagent. In *Methods in Enzymology*; Packer, L., Ed.; Academic Press: Cambridge, MA, USA, 1999; Volume 299, pp. 152–178.
51. Alara, O.R.; Abdurahman, N.H.; Mudalip, S.K.A.; Olalere, O.A. Characterization and Effect of Extraction Solvents on the Yield and Total Phenolic Content from Vernonia Amygdalina Leaves. *J. Food Meas. Charact.* **2018**, *12*, 311–316. [CrossRef]
52. Re, R.; Pellegrini, N.; Proteggente, A.; Pannala, A.; Yang, M.; Rice-Evans, C. Antioxidant Activity Applying an Improved ABTS Radical Cation Decolorization Assay. *Free Radic. Biol. Med.* **1999**, *26*, 1231–1237. [CrossRef]
53. Brand-Williams, W.; Cuvelier, M.E.; Berset, C. Use of a Free Radical Method to Evaluate Antioxidant Activity. *LWT—Food Sci. Technol.* **1995**, *28*, 25–30. [CrossRef]

molecules

MDPI

Article

Metabolomic Profiling, Box–Behnken Design-Based Optimization of Ultrasonic Extraction, and Skin Anti-Aging Potential of the Green Husk of *Juglans regia* L. as a Sustainable Natural Waste

Sıla Özlem Şener [1,*], Sabita Shaha [2], Sahar Sadigh Barazandeh [1], Ömer Şen [3], Engin Koçak [4], Tuğba Subaş [5], Şerife Nur Kıraç [1] and Emirhan Nemutlu [6]

1 Department of Pharmacognosy, Gulhane Faculty of Pharmacy, University of Health Sciences, 06210 Ankara, Türkiye; sadighshr@gmail.com (S.S.B.); sennur4241@gmail.com (Ş.N.K.)
2 School of Pharmacy and Medical Sciences, University of Bradford, Bradford BD7 1AZ, UK; s.shaha2@bradford.ac.uk
3 Department of Basic Sciences and Health, Hemp Research Institute, Yozgat Bozok University, 66900 Yozgat, Türkiye; omer.sen@bozok.edu.tr
4 Department of Analytical Chemistry, Gulhane Faculty of Pharmacy, University of Health Sciences, 06210 Ankara, Türkiye; engin.kocak@sbu.edu.tr
5 Department of Pharmacognosy, Faculty of Pharmacy, Karadeniz Technical University, 61080 Trabzon, Türkiye; tugbasubas@ktu.edu.tr
6 Department of Analytical Chemistry, Faculty of Pharmacy, Hacettepe University, 06800 Ankara, Türkiye; enemutlu@hacettepe.edu.tr
* Correspondence: silaozlem.sener@sbu.edu.tr

Abstract: The green husk of *Juglans regia* L. is rich in bioactive phytochemicals and exhibits various biological activities. This study aimed to investigate the skin anti-aging potential of the green husk of *J. regia* by determining the optimal extraction conditions using a Box–Behnken Design (BBD), targeting elastase inhibition, and by correlating variations in phenolic compounds identified through metabolomic analyses with changes in the extraction conditions. Ultrasonic-assisted extraction was employed along with natural deep eutectic solvents (NADES). Three levels of three independent variables (NADES/H_2O ratio, temperature, and extraction time) were incorporated into the BBD. Phenolic compounds were determined semi-quantitatively using liquid chromatography–quadrupole time-of-flight mass spectrometry (LC q TOF/MS), while elastase inhibition was evaluated by spectroscopic methods. Quadratic response surface models were proposed based on the BBD model adequacy test using multiple regression analysis. It was determined that the optimum conditions for maximizing phenolic content and elastase inhibition were 99.79% ethanol, 51.30 °C temperature, and 49.58 min, in which case the desirability score was 1. Metabolomic analysis identified 41 phenolic compounds across 27 ontological groups, with 24 compounds showing a semi-quantitative increase. Consequently, the waste green husk of *J. regia* demonstrated enhanced anti-aging potential due to the higher content and diversity of phenolic compounds.

Keywords: walnut; green extraction; phenolic compounds; elastase inhibition; waste valorization

1. Introduction

The World Health Organization (WHO) defines healthy aging as the ability to develop and maintain functional capacity healthily despite advancing age [1]. Skin aging, as a

 https://doi.org/10.3390/molecules30214191

component of aging, results from changes that occur at the organ, tissue, and cellular levels. Unlike other age-related changes, skin aging is visibly apparent, which can have a more significant impact on human psychology, social relationships, and overall health [2,3]. The global market for anti-aging products has been expanding rapidly. In 2020, the market size in the United States of America (USA) was estimated at 34.2 billion dollars, with predictions indicating it will reach 47.8 billion dollars by 2027. China is expected to contribute 18.4 billion dollars to the global market by 2027. Other countries with notable shares in this market, although smaller in comparison, include Canada, Japan, Europe, and Germany [4].

Elastase is a protease that plays a significant role in skin aging, being a component of the chymotrypsin family. This enzyme is capable of degrading skin proteins such as elastin, collagen, and fibronectin [5]. Factors such as age and chronic exposure to ultraviolet (UV) radiation lead to increased elastase activity, which in turn exacerbates the degeneration of elastic fibers within the skin tissue. As this degeneration progresses, the skin loses its elasticity, resulting in the formation of wrinkles and sagging [6]. Consequently, there is a therapeutic interest in the use of elastase inhibitors for addressing skin aging [7].

In recent years, natural resources that have been classified as agricultural waste are gaining increasing significance due to their secondary metabolite content, which is associated with physiological activities. Their abundant availability and potential to contribute to environmental pollution when discarded without regulation have made them a focal point of numerous research studies. The demand for the utilization of naturally sourced agricultural waste in nutrition, food, cosmetics, pharmaceuticals, and industrial applications is rapidly increasing. One of the natural resources utilized in the generation of agricultural waste is the fruit of *Juglans regia* L. The outer green thick layer of *J. regia*, referred to as the green husk, is a significant agricultural by-product generated during the harvesting and processing of the fruit. This waste material has the potential to contribute to environmental pollution [8]. Traditionally, the green husks of *J.* regia have been reported to be utilized in the treatment of various ailments, including skin diseases, warts, stomach pain, and anemia [9,10]. The green husk of *J. regia* is particularly rich in phytochemical constituents, especially phenolic compounds. The primary chemical components of the green husk include naphthoquinones and their glycosides, diarylheptanoids, terpenes and their glycosides, flavonoids, sterols, and polyphenols. Furthermore, the green husk of *J. regia* has demonstrated antitumor, antioxidant, antiviral, and bacteriostatic activities [8,11–13]. It is important to use appropriate and environmentally friendly extraction methods to efficiently recover the bioactive compounds of *J. regia* green husks.

In contemporary research, the significance of green extraction techniques is increasingly recognized, as they aim to enhance extraction efficiency and quality by reducing extraction time, global energy consumption, solvent usage, environmental impact, economic costs, and waste production [14]. The fundamental principle of ultrasonic extraction, one of the green extraction methods, involves the mechanical, cavitational, and thermal effects that lead to the disruption of cell walls, resulting in smaller particle sizes and an increased mass transfer across cell membranes. The advantages of ultrasonic extraction include its safety and efficiency, rapid extraction kinetics, and the ability to extract heat-sensitive compounds [14–17]. The various organic solvents employed during extraction processes pose detrimental effects on both human health and the environment. This issue underscores the increasing significance of "green solvents" that can replace organic solvents derived from fossil fuels. Ionic liquids (ILs) and deep eutectic solvents (DES) are primary types of solvents classified as green solvents. ILs and DESs are frequently described as green solvent alternatives; however, their actual sustainability is still debated due to differences in toxicity, biodegradability, and environmental persistence. To enhance the safety of green solvents, natural deep eutectic solvents (NADES) have been developed by substituting the components in DESs with natural products that function as hydrogen bond donors (HBD). Numerous studies have demonstrated that NADES can effectively extract active compounds from a wide range of natural sources. The advantages of utilizing NADESs include ease of application, relatively low costs, ready-to-use formulations, and biodegradability [18].

Optimization studies aimed at determining the most suitable extraction conditions from plant sources with reduced time and energy expenditure have gained significant importance in contemporary applications. One of the frequently employed statistical optimization methods that incorporates both dependent and independent variables is the response surface methodology. This methodology utilizes a combination of statistical and mathematical techniques to derive the best statistical estimates from a limited set of experimental data [19]. Among the techniques utilized within response surface methodologies, the Box–Behnken design (BBD) is a second-order multivariate technique based on three-level factorial designs, which is employed to identify optimal experimental conditions. The number of experiments required for the development of the Box–Behnken technique (N) is defined by the equation $N = 2k\ (k - 1) + C0$, where k represents the number of factors and (C0) denotes the number of repetitions at the center point. By leveraging the Box–Behnken technique, it is possible to determine the optimal extraction conditions through statistical analysis using a limited set of experimental data [20,21].

Metabolites are low-molecular-weight organic molecules that participate in diverse biochemical reactions and represent intermediate or end-products of various cellular processes. Metabolomics has emerged as a pivotal discipline that investigates the complete set of metabolites within biological systems through an integrative and systems-based approach. One of the major applications of metabolomic studies is the discovery of novel bioactive molecules or drug candidates derived from natural sources. Beyond identifying such compounds, metabolomic analyses also provide insights into the interactions between specific metabolites-such as phenolic constituents-and biological systems. These analyses enable the comprehensive characterization of metabolites involved in these interactions and elucidate the underlying biochemical mechanisms [22]. Furthermore, advanced correlation analyses following metabolomic profiling allow the establishment of relationships between metabolic signatures and specific biological activities, offering a modern strategy to identify compounds responsible for the observed effects. The regression coefficients obtained from these correlation models indicate whether the associations between metabolites and activities are positive or negative—reflecting activating or inhibitory effects—and their magnitude provides an estimate of the strength of these interactions [23,24].

Phenolic compounds possess the potential to be utilized in the prevention and treatment of visible signs of both early and chronological aging, as well as age-related skin disorders, through various mechanisms in which they act as elastase inhibitors [25].

Skin aging is closely associated with oxidative stress, inflammation, and the degradation of structural proteins such as collagen and elastin, processes in which elastase plays a critical enzymatic role. In this context, the development of sustainable extraction strategies for obtaining phenolic-rich bioactives from natural resources represents a promising avenue for anti-aging applications. Therefore, this study aims to explore the potential of biological waste from *J. regia* green husks by preparing eco-friendly extracts rich in phenolic constituents and evaluating their relevance to skin-aging mechanisms. To this end, green husks of *J. regia* were utilized to formulate a green solvent, which was then applied in ultrasound-assisted extraction (UAE) to obtain environmentally sustainable extracts. The elastase inhibitory effects of these extracts were determined in vitro using spectroscopic techniques, with inhibitory potency expressed as half maximal inhibitory concentration (IC_{50}). Optimization of the green extraction process was subsequently performed to identify the most effective extraction parameters for maximizing bioactivity while minimizing energy and time consumption. Furthermore, a metabolomic analysis was conducted to elucidate the influence of extraction conditions on phenolic composition and to establish the relationship between the identified phenolic metabolites and elastase inhibition. For the metabolomic analyses, three distinct extracts were evaluated. The first was the BBD-optimized extract (U1), obtained under the optimal extraction conditions determined by the BBD model. The second, the maceration extract (M1), was prepared using the conventional maceration technique to allow comparison between traditional and modern extraction approaches. The third extract, the optimized extract at maceration temperature (U2), was produced under the same optimized conditions as U1, except that the extraction temperature was adjusted to that of maceration, in order to specifically assess the effect of temperature on extraction efficiency. This integrative approach provides both a scientific and practical framework for the valorization of *J. regia* residues in sustainable anti-aging formulations.

2. Results and Discussion

2.1. Results of Elastase Inhibition

The analysis of IC_{50} values revealed that the coefficient of determination (R^2) of the dose–response curves constructed from log concentration versus percentage inhibition data exceeded 0.99, confirming the reliability and consistency of the measurements. Based on the 17 experimental runs established by the BBD, elastase inhibition values varied considerably, with IC_{50} values ranging from 1.16 μg/mL to 45.82 μg/mL (Table 1). These results indicate that extraction parameters had a measurable influence on enzyme inhibition.

Table 1. Comparison of Experimental Data and Predictive Data Obtained by BBD.

	A	B	C	R1 (Elastase Inhibition)		
	%	°C	min.	Exp. (μg/mL)	Pre. (μg/mL)	PRE %
1	0	−1	−1	3.39	4.70	27.87
2	0	0	0	9.99	9.83	−1.63
3	0	0	0	9.96	9.83	−1.32
4	+1	−1	0	1.78	0.39	−346.34
5	−1	−1	0	19.61	21.24	7.67
6	−1	0	+1	18.06	17.99	−0.39
7	+1	0	−1	1.41	1.48	4.73
8	0	0	0	9.62	9.83	2.14
9	0	−1	+1	13.68	12.12	−12.87
10	0	0	0	9.72	9.83	1.12
11	0	+1	−1	40.85	42.41	3.68
12	−1	0	−1	32.67	29.73	−9.89
13	+1	0	+1	1.16	4.10	71.71
14	−1	+1	0	45.82	47.20	2.92
15	0	+1	+1	27.19	25.88	−5.06
16	+1	+1	0	27.52	25.89	−6.30
17	0	0	0	9.86	9.83	−0.31

Abbreviations: Exp.: experimental, Pre.: predictive, PRE: percent relative error, A: NADES/H_2O ratio ((*w*/*w*); −1 = 25.00%, 0 = 62.50%, +1 = 100.00%); B: Temperature (°C; −1 = 40.00, 0 = 60.00, +1 = 80.00); C: Extraction time (min; −1 = 20.00, 0 = 40.00, +1 = 60.00).

The relationships between the extraction variables (NADES ratio, temperature, and time) and elastase inhibition were subsequently evaluated using the BBD (Section 2.2), followed by optimization studies presented in Section 2.4.

2.2. Analysis of Box–Behnken Design

BBD is a response surface methodology employed to optimize experimental variables utilizing three-level factorial designs [20]. Three independent variable levels, NADES ratio, temperature, and time, were employed to facilitate the BBD. The BBD methodology serves to elucidate the relationship between experimental and predictive outcomes [26]. One metric employed to assess this significance is the percent relative error, which is determined using the formula [((predictive data–experimental data)/predictive data) × 100] [27]. The experimental design of the BBD, along with the experimental data, predictive data, and the percent relative error values for all responses, is detailed in Table 1.

The analysis employed a sequential model sum of squares, model summary, and lack of fit tests across various model types, including quadratic, cubic, two-factor interaction (2FI), and linear models, to identify the most suitable model for each response and derive the corresponding regression equations through analysis of variance (ANOVA). Additionally, the assessment of model reliability was conducted using the coefficient of variance (CV%), R^2, and the adjusted coefficient of determination (R^2a) [28]. The findings indicated that the quadratic model was the most appropriate, demonstrating a significant *p*-value ($p < 0.0001$), as supported by the analysis of the sum of squares and the lack of fit test. In addition, R^2 was used to evaluate the reliability of the model. The R^2 value of 0.9886, being close to 1, supports the significance of the quadratic model (Table S1).

Table 2 presents the coefficient estimates and *p*-values for the response of elastase inhibition. The inhibition of elastase was found to be significantly influenced by the linear coefficients (A, B, and C), two quadratic coefficients (A^2 and B^2), and two interaction coefficients (AC and BC) ($p< 0.05$) (Table 2). The most substantial impact on elastase inhibition, as indicated by the IC_{50} value, was attributed to temperature (B), which exhibited a regression coefficient of 12.87 and a highly significant *p*-value ($p < 0.0001$). An increase in the NADES/H_2O ratio (A) corresponded to a reduction in the IC_{50} value for elastase inhibition with a regression coefficient of −10.14 and a *p*-value of less than 0.0001. Conversely, time (C) exhibited an inverse relationship, with a regression coefficient of −2.28 and a *p*-value of 0.0233. Additionally, the two quadratic terms A^2 ($p = 0.0300$) and B^2 ($p < 0.0001$) showed significant effects, both having a direct proportional influence with regression coefficients of 2.95 and 10.90, respectively. The interaction term (AC) also had a significant direct effect on the IC_{50} value of elastase inhibition ($p = 0.0147$, regression coefficient = 3.59), as well as another interaction term (BC) with a regression coefficient of −5.99 and *p*-value of 0.001. In contrast, the quadratic terms (C^2) and interaction terms (AB) did not demonstrate significant effects, as indicated by *p*-values greater than 0.05.

Table 2. *p*-values and coefficient estimates of one response.

	Elastase Inhibition				
	***p* Value**	**Coefficient Estimate**	**Standard Error**	**95% CI Low**	**95% CI High**
Model	<0.0001				
Intercept		9.83	0.9973	7.47	12.19
A-Solvent Ratio	<0.0001	−10.54	0.7884	−12.40	−8.67
B-Temperature	<0.0001	12.87	0.7884	11.00	14.73
C-Time	0.0233	−2.28	0.7884	−4.14	−0.4144
AB	0.9190	−0.1175	1.12	−2.75	2.52
AC	0.0147	3.59	1.12	0.9534	6.23
BC	0.0010	−5.99	1.12	−8.62	−3.35
A^2	0.0300	2.95	1.09	0.3801	5.52
B^2	<0.0001	10.90	1.09	8.33	13.47
C^2	0.6314	0.5450	1.09	−2.02	3.11
Residual					
Lack of fit	<0.0001				

Consequently, following the removal of the insignificant terms, the regression equation was revised, resulting in the subsequent reduced Equation (1) for the IC_{50} value of elastase inhibition.

$$\text{Elastase inhibition (IC}_{50}\text{ value)} = 9.83 - 10.54A + 12.87B - 2.28C + 3.59AC - 5.99BC + 2.95A^2 + 10.90B^2 \quad (1)$$

The three-dimensional response surface plots provide a compelling visual representation of regression equations, illustrating the interplay between independent variables and the response variable [29]. As depicted in Figure 1, an increase in the NADES/H_2O ratio resulted in a decrease in the IC_{50} value for elastase inhibition. Extraction durations between 40 and 60 min caused a significant increase in elastase inhibition. Furthermore, maintaining the extraction temperature within the range of 40 °C to 60 °C significantly lowered the IC_{50} value, thereby maximizing elastase inhibition.

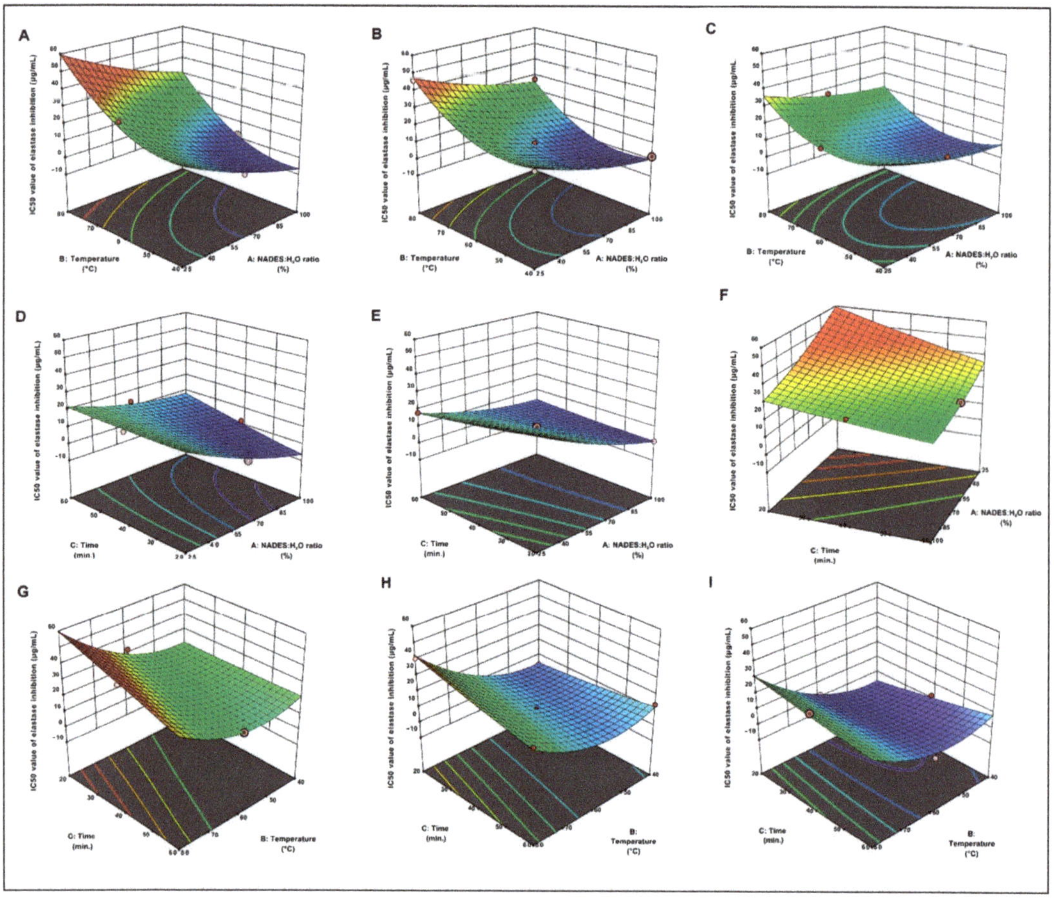

Figure 1. Three-dimensional Response Surface Graphs for IC_{50} Values of Elastase Inhibition Effect of Temperature: The relationship between the solvent ratio and different extraction times for the IC_{50} value of elastase inhibition: 20 min (**A**), 40 min (**B**), and 60 min (**C**). Effect of Time: The relationship between the solvent ratio and different temperatures for the IC_{50} value of elastase inhibition: 40 °C (**D**), 60 °C (**E**), and 80 °C (**F**). Effects of NADESs/H_2O ratio: The relationship between the time and different temperatures for the IC_{50} value of elastase inhibition: 25% (**H**), 62.5% (**G**), and 100% (**I**).

2.2.1. Effect of Extraction Temperature

Temperature is one of the most influential parameters governing the UAE, as it modulates cavitation behavior, solvent mobility, and the integrity of heat-sensitive bioactive compounds. A moderate increase in temperature enhances solvent diffusivity and softens plant tissue structures, thereby facilitating solute migration and improving extraction efficiency. In contrast, excessive heating can weaken cavitation collapse and accelerate the degradation of phenolic constituents, resulting in lower extraction yields and diminished bioactivity. Maintaining the extraction temperature within an intermediate range of approximately 40–60 °C has been shown to provide an effective compromise between enhanced mass transfer and preservation of molecular stability, ensuring sufficient energy for solvent permeation without compromising the structural integrity of active compounds [30,31].

Similar thermal patterns have been documented in various botanical matrices. For instance, *Benincasa hispida* seed oil extraction demonstrated an increase in yield with temperature up to about 52 °C, followed by a decline due to reduced cavitation efficiency. Under these optimized ultrasonic conditions, the extract exhibited considerably higher total phenolic content and antioxidant activity compared to the Soxhlet extract, highlighting that moderate UAE conditions improve both yield and functional compound recovery [32].

In agreement with these observations, the optimum extraction temperature obtained in this work (51.30 °C) falls well within the recommended thermal range, confirming that moderate heating promotes efficient cavitation and solvent diffusion without inducing thermal degradation. At this point, the extracts showed the highest total phenolic concentration and the strongest elastase inhibitory response, suggesting that phenolic compounds play a principal role in the anti-aging efficacy of the optimized extract. Similarly, *Peucedanum ostruthium* leaf extraction performed at 61 °C achieved maximum total phenolic (145.82 mg gallic acid equivalent-GAE/g dry weight-dw) and flavonoid (637.45 mg rutin equivalent-RE/g dw) levels, along with enhanced elastase inhibition [31].

Taken together, maintaining the extraction temperature within an intermediate range of temperature ensures an optimal equilibrium between ultrasonic mechanical disruption and phenolic compound stability, thereby maximizing both extraction yield and elastase-inhibitory potential.

2.2.2. Effect of Extraction Time

The duration of UAE is a critical variable that governs both the rate of mass transfer of bioactive constituents and the intensity of cavitation occurring within the extraction medium. During the early stages of sonication, the formation and implosion of cavitation bubbles generate microjets that rupture plant cell walls, thereby facilitating solvent penetration and enhancing the transport of intracellular compounds. As extraction progresses, the diffusion of cellular metabolites into the solvent phase increases, resulting in a substantial improvement in extraction efficiency. However, when sonication exceeds the optimal point, localized heating and the accumulation of reactive radicals may occur, leading to oxidative degradation and conformational changes in thermolabile molecules. Hence, maintaining an intermediate extraction duration ensures a balanced environment that supports efficient mass transfer while preserving molecular stability [30,33].

Empirical evidence from diverse biological matrices indicates that UAE efficiency strongly depends on the optimization of both temperature and duration parameters. Moderate extraction times have been shown to enable complete cell disruption and effective solute diffusion without promoting chemical degradation. For instance, *Astragalus membranaceus* exhibited its highest flavonoid recovery at approximately 50 min and 58 °C, confirming the benefits of intermediate sonication periods [34]. Similarly, *Boletus bicolor* achieved maximal polyphenol extraction at 41 min and 40 °C [35]. In *Crocus sativus* by-products, extraction at 45 min and 45 °C produced the greatest phenolic content and antioxidant potential, whereas longer durations reduced both parameters [36]. The optimized extraction of polysaccharides from *Ficus microcarpa* aerial roots occurred at 49 min and 74 °C [37]. Likewise, fucoidan extraction from *Sargassum* species reached maximum antioxidant activity at 49 min and 50 °C, followed by a decline with prolonged exposure [38]. Additionally, the ultrasound-assisted recovery of flavonolignans from *Silybum marianum* (L.) Gaertn. Fruits were most efficient at 60 min and 45 °C, exhibiting strong collagenase and elastase inhibitory activity [39]. Within our optimized temperature range (40–60 °C), sonication durations of approximately 40–50 min appear to represent the most favorable conditions for maximizing the recovery of bioactive compounds from plant matrices while minimizing degradation.

Further studies have reported comparable tendencies. *Eisenia arborea* extracts displayed similar antioxidant activity at 60 and 90 min, yet elastase inhibition was markedly higher at 90 min, indicating that prolonged sonication can enhance the liberation of enzyme-inhibitory metabolites [40]. In contrast, *Phoenix dactylifera* L. leaves exhibited the highest phenolic and flavonoid yields, and the strongest, after a much shorter UAE duration of 29 min at 50 °C [41]. In another study, *Tamus communis* L. by-products extracted for 40 min at 70 °C exhibited significant increases in flavonoids and phenolic acids, leading to enhanced inhibition of skin-aging enzymes [42]. The influence of extraction time on elastase inhibition appears to vary depending on the plant matrix. This variability suggests that the effect is not solely time-dependent but also influenced by other extraction parameters, particularly temperature. Under extraction temperatures comparable to those applied in our experiments, the effective duration for maximizing elastase inhibition generally ranged between 40 and 60 min, which closely aligns with the extraction time determined in the present study. This observation is consistent with predictions from response surface methodology (RSM), confirming that moderate extraction durations provide an optimal balance between mass transfer efficiency and the preservation of thermolabile bioactives.

The optimized sonication time of 49.58 min identified in this study, in combination with metabolomic profiling, revealed an increased abundance of numerous bioactive metabolites, particularly phenolic derivatives. Metabolomic analyses further demonstrated that this extraction period enhanced the accumulation of phenolic metabolites and produced extracts characterized by greater chemical complexity and biological activity. Phenolic compounds are well established as potent secondary metabolites with significant elastase inhibitory capacity-a key mechanism underlying anti-aging effects. Therefore, the elevated levels of phenolic and other bioactive metabolites obtained under these optimized conditions likely contributed directly to the observed elastase inhibition. Collectively, these findings reinforce the general consensus that moderate ultrasonic durations facilitate efficient solvent–matrix interactions and cell disruption while preserving the structural and functional integrity of bioactive molecules.

2.2.3. Effect of NADES/H_2O Ratio

The deliberate incorporation of a controlled amount of water into the NADES system was intended to fine-tune its physicochemical behavior and improve extraction performance. Due to their dense hydrogen-bond network, NADESs typically exhibit high viscosity, which can impede solvent diffusion and mass transfer processes, thereby limiting the release of intracellular phytochemicals. The introduction of water partially weakens the hydrogen-bonding interactions between HBDs and hydrogen bond acceptors (HBAs), leading to a marked reduction in viscosity and an enhancement in solvent fluidity, which collectively facilitate deeper penetration of the solvent into the plant tissue [43,44]. In addition, a moderate water fraction promotes swelling of the lignocellulosic framework and expands the solvent's polarity range, enabling improved solubilization of both hydrophilic and moderately lipophilic metabolites [45,46]. Conversely, when the water proportion exceeds a critical level, the supramolecular organization of NADES begins to collapse, compromising its characteristic solvating potential and disrupting the cooperative hydrogen-bond structure [47]. Therefore, maintaining an optimal water content is crucial for balancing viscosity reduction and solvent integrity, ensuring the highest extraction efficiency while preserving the distinctive hydrogen-bonding environment that defines NADES functionality. As previously reported, moderate water addition can decrease viscosity and enhance mass transfer, which may initially improve the extraction of especially phenolic compounds [43,44,46]. However, beyond a critical threshold of approximately 30–50% (w/w) H_2O, the supramolecular hydrogen-bond structure of NADES begins to break down, reducing the solvent's ability to dissolve phenolic acids and flavonoids [47–51]. Although water additions up to broad ranges (even approaching 95%) have been reported in some studies, such systems are generally considered to behave as aqueous solutions rather than true eutectic mixtures [46–51].

The solvent system containing a 100% NADES/H_2O ratio achieved the highest elastase inhibitory response, indicating an optimal balance between solvation strength and structural stability that facilitated the efficient solubilization of phenolic compounds. In contrast, partially hydrated formulations (25% and 62.5% NADES/H_2O) exhibited markedly lower inhibition, likely due to disruption of the hydrogen-bonding network within the choline chloride–glycerol matrix. The excessive water fraction weakened solvent–solute interactions and reduced extraction selectivity. These findings highlight the importance of maintaining a predominantly NADES-based composition while avoiding excessive water incorporation to preserve the cooperative hydrogen-bond structure and maximize elastase inhibition efficiency. Nevertheless, future studies employing a wider range of NADES/H_2O ratios or alternative NADES compositions could provide a more comprehensive understanding of how water modulates solvent polarity, viscosity, and hydrogen-bond strength. Such investigations would allow the precise identification of the optimum water content that maximizes both extraction efficiency and biological activity across different plant matrices.

2.3. *Validation of the Box–Behnken Design's Accuracy*

To ensure that the fitted model provides a reliable approximation of the actual values, it is essential to conduct a validation process [52]. The accuracy of the BBD was assessed through a comparison of diagnostic plots representing both the experimental and predicted model outcomes. As illustrated in Figure S1, these diagnostic plots demonstrated the model's adequacy and the correlation between the experimental results and the predicted values concerning elastase inhibition. The proximity of the data points in the diagnostic plots to the reference straight line indicates a satisfactory fit between the experimental and model data. Consequently, the findings affirm that the Box–Behnken model is effective for evaluating elastase inhibition.

2.4. *Optimisation of Extraction Procedures*

The desirability score serves as a critical metric for evaluating optimal conditions. A desirability score approaching 1 enhances the precision of the assessed conditions [53]. The proposed model delineates optimal conditions aimed at minimizing the desirability for elastase inhibition, thereby formalizing the model equation. The desirability scores recorded were 1 for elastase inhibition (Figure 2A). The optimal parameters identified for achieving the highest elastase inhibition effect included a NADES/H_2O ratio of 99.79%, an extraction temperature of 51.30 °C, and an extraction duration of 49.58 min (Figure 2B).

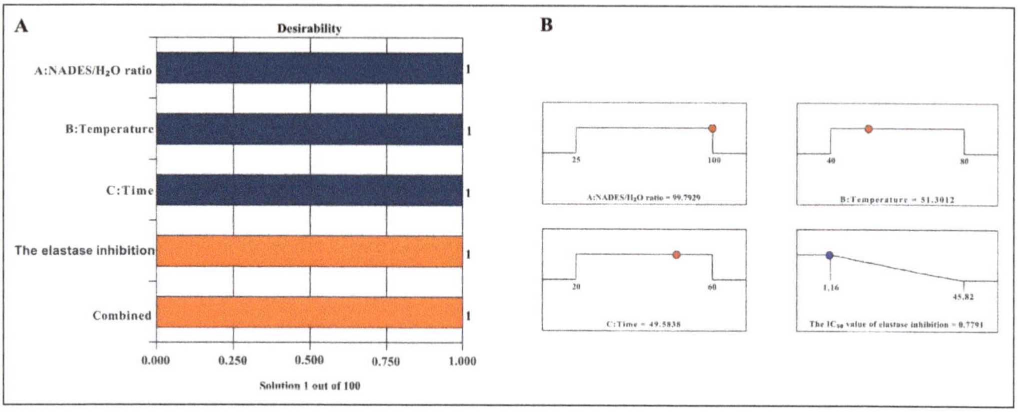

Figure 2. Desirability bar (**A**) and ramp for optimising extraction conditions (**B**). The blue bars represent the independent variables (A: NADES/H_2O ratio, B: Temperature, and C: Time), while the orange bars indicate the dependent variables (Elastase inhibition and Combined response).

It was determined that the experimental IC_{50} value of elastase inhibition (0.7735 ± 0.0182 μg/mL) and the predicted IC_{50} value of elastase inhibition (0.7791 ± 0.0275 μg/mL) for U1 were in close agreement. Additionally, the experimental and predicted values under other extraction conditions were found to correlate in terms of elastase inhibition. The IC_{50} value of U2 was observed to be 1.8213 ± 0.3173 μg/mL, whereas that of M1 was determined as 12.6377 ± 0.5275 μg/mL (Table 3).

Table 3. Predicted and experimental values of optimal conditions and comparison with the conventional extraction technique (maceration).

U1		U2	M1
NADES/H_2O ratio (w/w) → 99.79% Temperature (°C) → 51.30 Time (min.) → 49.58		NADES/H_2O ratio (w/w) →99.79% Temperature (°C) → 25 Time (min.) → 49.58	NADES/H_2O ratio (w/w) → 99.79% Temperature (°C) → 25 Time (min.) → 24 h
IC_{50} values of elastase inhibition			
Predicted	Experimental	Experimental	Experimental
0.7791 ± 0.0275 μg/mL	0.7735 ± 0.0182 μg/mL	1.8213 ± 0.3173 μg/mL	12.6377 ± 0.5275 μg/mL

U1: The BBD-optimized extract, U2: The optimized extract at maceration temperature, M1: The maceration extract.

2.5. Results of Metabolomics Analysis

Metabolomic analyses revealed the presence of 90 distinct compounds belonging to 58 different chemical classes in the samples. Among these compounds, it was identified that 41 are phenolic components classified into 27 different chemical categories (Table S2, Figure S2).

The Principal Component analysis (PCA) score plot illustrates the distribution of metabolite profiles obtained from three different extraction methods (U1, U2, and M1) (Figure 3). The first two principal components (PC1 and PC2) together explain 64.5% of the total variance in the dataset (30.9% and 33.6%, respectively). Samples from each extraction method form partially overlapping clusters, indicating that the methods yield comparable overall metabolite profiles. However, slight separation between groups suggests subtle differences in extraction efficiency and compound selectivity. The proximity of U1 and U2 clusters implies that these two methods isolate metabolites with similar polarity and chemical characteristics, whereas M1 shows a slightly broader distribution, reflecting a more diverse extraction pattern. Overall, the PCA results demonstrate that all three methods produce reproducible data, with U1 and U2 showing higher similarity in metabolite composition (Figure 3).

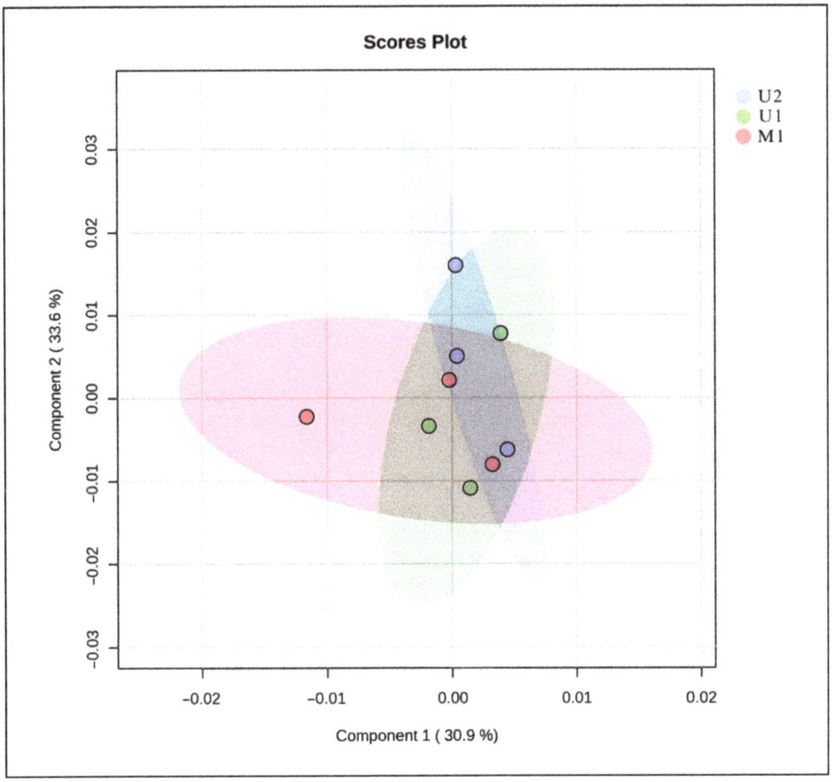

Figure 3. PCA of the metabolome structure of different extraction techniques. U1: The BBD-optimized extract, U2: The optimized extract at maceration temperature, M1: The maceration extract.

The objective of the correlation analyses was to investigate the impact of extraction conditions on the phytochemical profile. In this context, significant increases or decreases were observed in 88 out of 90 compounds analyzed across different extract groups (Table S2).

The extraction conditions of U1 have been found to induce significant changes in 77 compounds when compared to the extraction conditions of M1. Specifically, the U1 extraction conditions resulted in a significant increase in 45 of these compounds, while a decrease was observed in 32 compounds. Among the 45 compounds, 24 were identified as having phenolic characteristics. In comparison to the total number of compounds, these extraction conditions led to a 26.67% increase in phenolic metabolites (Table S2, Figure 4A).

When U2 extraction conditions were compared with M1 extraction conditions, it was determined that they caused significant changes in 66 compounds. An increase was observed in 39 of these compounds, while a decrease was observed in 27. Of the 39 compounds with an increase, 27 were found to be phenolic in character. When compared to the total number of compounds under these extraction conditions, an increase of 26.67% was observed in the number of compounds with phenolic character (Table S2, Figure 4B).

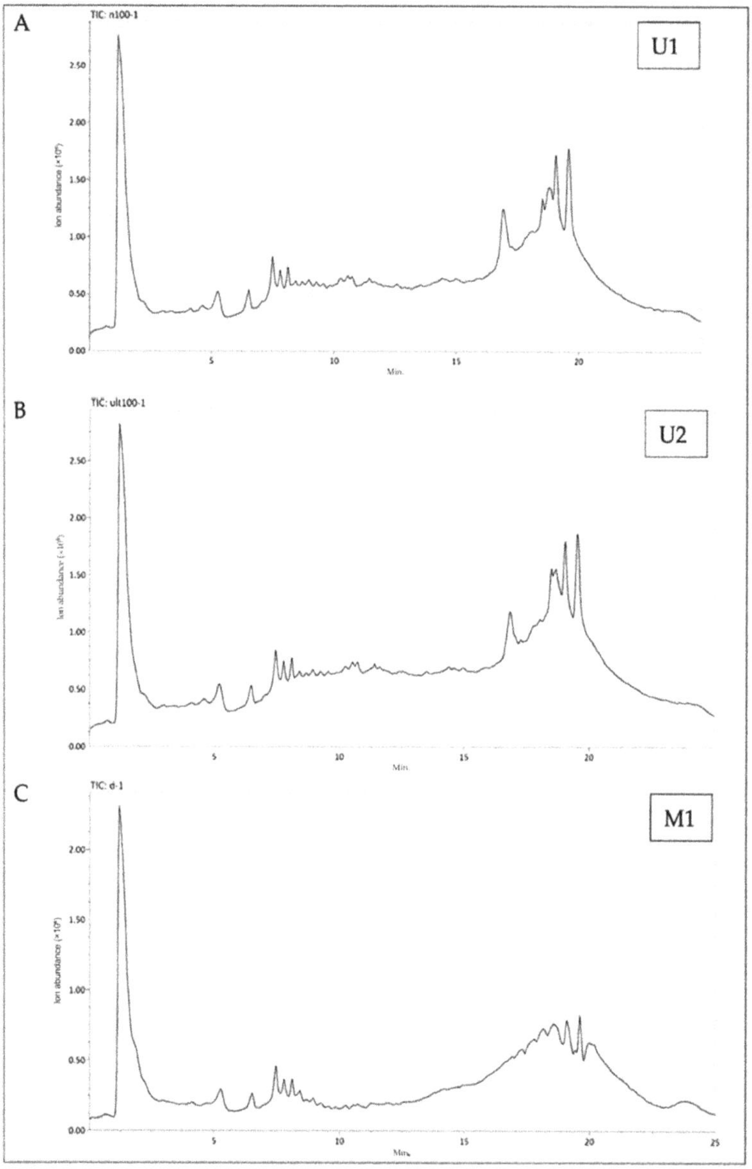

Figure 4. Representative LC-Q-TOF/MS chromatograms of samples. (**A**) The LC-Q-TOF/MS chromatogram of U1 (The BBD-optimized extract); (**B**) The LC-Q-TOF/MS chromatogram of U2 (The optimized extract at maceration temperature); (**C**) The LC-Q-TOF/MS chromatogram of M1 (The maceration extract).

The extraction conditions of U2 have been found to cause significant changes in 66 compounds when compared to the extraction conditions of M1. Among these compounds, an increase was observed in 39, while a decrease was noted in 27. It was determined that 27 of the 39 compounds exhibiting an increase possess phenolic characteristics. When compared to the total number of compounds under these extraction conditions, there was a 25.56% increase in the number of phenolic compounds (Table S2, Figure 4B).

When comparing the extraction conditions U1 and U2, it was observed that there were fewer changes in comparison to the maceration technique. A total of 23 compounds exhibited significant alterations. Among these compounds, 13 showed an increase under the U1 extraction conditions, with 6 of them identified as having phenolic characteristics. The U1 extraction conditions resulted in a 6.67% increase in the number of phenolic compounds relative to the total number of compounds (Table S2, Figure 4C).

It has been determined that neobonaspectin A, characterized by its sesquilignan structure, exhibits the highest increase in concentration when the extraction conditions are modified. The log(FC) values were established as 8 for U1/M1, −10 for M1/U2, and −2 for U2/U1 (Table S2). This indicates that the extraction conditions U1 and U2 significantly enhance the levels of neobonaspectin A.

It has been observed that the maceration technique is more effective for long-chain fatty acids. In the case of ultrasonic extraction, the similarity in temperature to the maceration method has resulted in an increased fatty acid ratio. However, a different scenario has been noted for the unsaturated fatty acid, linoleic acid. Under U2 extraction conditions, linoleic acid achieves the highest concentration, whereas the extract prepared using the maceration technique remains at a lower level (Table S2).

When comparing the extraction conditions of U1 with M1, it was observed that the compounds exhibiting the highest degree of variation after neobonaspectin A were, in order, phenolic herniarin, β-glucogallin, methyl gallate, and gallic acid. In the U2 extraction technique, this order was identified as herniarin, 2,5-dihydroxybenzoic acid, methyl gallate, and gallic acid. Furthermore, when the extraction conditions of U1 and U2 were compared, the compounds were identified as phelligridin I, D-quinic acid, and 2,5–dihydroxybenzoic acid (Table S2).

This study demonstrated that the UAE method produced substantially higher concentrations of several phenolic constituents compared to the conventional maceration process. Among the compounds that showed increased abundance, neubonaspectin A, herniarin, β-glucogallin, methyl gallate, and gallic acid were particularly prominent. The in silico molecular docking analysis revealed that herniarin, a coumarin derivative identified in *Mikania* sp., exhibits a notable interaction with the Ser195 residue located in the catalytic site of human neutrophil elastase (HNE). The calculated binding energy (−24.89 kcal/mol) of herniarin was comparable to that of other elastase-interacting phytochemicals such as psoralen and patuletin. The consistent engagement with the Ser195 residue suggests that herniarin may act as a moderate elastase modulator [54]. The activity-guided fractionation of the ethyl acetate extract derived from the pedicels and leaves of *Cotinus coggygria* Scop. yielded four bioactive constituents, among which methyl gallate demonstrated a moderate elastase inhibition (IC_{50} = 198.86 ± 0.84 µg/mL). This finding suggests that methyl gallate functions as a moderate yet biologically relevant elastase modulator, thereby contributing to the preservation of the extracellular matrix and underpinning the anti-aging potential of *C. coggygria* extracts [55]. The crude extract of white grape pomace, characterized by a high gallic acid content (596.36 ± 37.49 mg/kg dw), exhibited a marked ability to inhibit elastase activity in a dose-dependent manner, with an IC_{50} = 14.7 µg/mL [56]. To date, no scientific studies have been reported investigating the elastase inhibitory activity of either neubonaspectin A, β-glucogallin, or phelligridin I.

The compounds whose concentrations increased with the application of the UAE technique are known to exhibit anti-inflammatory, antioxidant, and epithelialization-promoting activities, thereby suggesting a potential contribution to anti-aging mechanisms. Certain compounds with a sesquilignan structure similar to neobonaspectin have been shown in nematode models to extend lifespan and enhance oxidative stress tolerance, potentially through the DAF-16-mediated signaling pathway [57]. Herniarin has been reported to suppress prostaglandin E2 (PGE2) release in a dose-dependent manner, reduce edema formation, and demonstrate that the substituent at the C-7 position plays a critical role in anti-inflammatory activity [58]. β-glucogallin, a potent aldose reductase inhibitor, has been shown to attenuate inflammatory responses and oxidative stress. β-glucogallin enhances skin barrier function by upregulating filaggrin and HAS3 expression in keratinocytes, and exerts anti-inflammatory effects by suppressing interleukin-4 (IL-4)/poly(I: C)-induced chemokines and pro-inflammatory cytokines [59,60]. Methyl gallate exerts its anti-aging potential primarily through reactive oxygen species (ROS) scavenging and activation of antioxidant pathways such as extracellular signal-regulated kinase 1/2 (ERK1/2)-nuclear factor erythroid 2-related factor 2 (Nrf2), and reduces the production of nitric oxide (NO) and IL-6 by macrophages [61]. Gallic acid inhibits pigmentation by downregulating melanogenic genes (tyrosinase, tyrosinase-related protein-1 (TRP-1), dopachrome tautomerase (Dct)) and suppressing microphthalmia-associated transcription factor (MITF) expression through inhibition of the cyclic adenosine monophosphate (cAMP)/protein kinase A (PKA)/cAMP-responsive binding protein (CREB) pathway, while activating ERK and AKT signaling, ultimately reducing melanin synthesis in cells, animal models, and zebrafish. These findings highlight its potential as an effective depigmenting agent for topical applications [62]. Phelligridin I, with its strong antioxidant properties, may play a role in modulating age-related cellular signaling changes [63]. Gentisic acid has been demonstrated to enhance keratinocyte proliferation by more than 20%, promote wound healing via ERK1/2 phosphorylation, and act as a safe tyrosinase inhibitor for the management of cutaneous hyperpigmentation disorders [64,65].

Although not as pronounced as the increases observed for the aforementioned compounds, several additional constituents that also exhibited elevated levels following extraction have been previously reported to possess elastase inhibitory activity in various scientific studies. Specifically, chlorogenic acid demonstrated significant elastase inhibition (IC_{50} = 203.3 μM). Molecular docking analysis revealed multiple hydrogen-bond interactions between chlorogenic acid and HNE residues Leu35, His40, Phe41, His57, and Gly193, resulting in a strong binding affinity. These molecular interactions support its potential role as a natural elastase regulator and highlight its possible contribution to anti-aging mechanisms [66]. Luteolin also exhibited potent inhibition of elastase release from human neutrophils, with an IC_{50} value of 6.91 ± 2.25 μM—comparable to that of the reference inhibitor LY294002 (IC_{50} = 4.97 ± 0.80 μM). This finding indicates that luteolin exerts a remarkable elastase-suppressing effect, likely contributing to its anti-inflammatory and dermoprotective properties through the attenuation of neutrophil-mediated proteolysis [67]. Ferulic acid showed measurable inhibitory effects on aging-associated enzymes, including elastase (IC_{50} = 75.61 μg/mL). This outcome suggests a moderate but functionally relevant elastase inhibition, implying that ferulic acid may help preserve skin elasticity by limiting elastin degradation, thereby enhancing its anti-aging potential [68]. Similarly, ellagic acid has been identified as a promising natural elastase inhibitor, exhibiting an IC_{50} value of 1.44 mg/mL. Spectroscopic analyses revealed that its interaction with elastase involves both static and dynamic quenching mechanisms, with complex formation at lower concentrations and dynamic processes predominating at higher levels. Thermodynamic evaluations further suggested that hydrogen bonding and van der Waals forces govern the interaction, leading to subtle conformational rearrangements in the enzyme structure [69]. Syringic acid also demonstrated a notable inhibitory effect on elastase activity (IC_{50} = 34.29 ± 1.71 μg/mL), indicating a moderate yet biologically relevant inhibition potential. This observation suggests that syringic acid may contribute to anti-aging and dermoprotective mechanisms by mitigating elastase-mediated extracellular matrix degradation [70].

The findings suggest that green extraction techniques not only provide environmental and economic advantages but also facilitate the recovery of bioactive phenolic compounds at concentrations biologically relevant for anti-aging applications. The optimization of extraction parameters enhances the correlation between phytochemical enrichment and biological efficacy, particularly in terms of enzyme inhibitory activities like elastase inhibition, which plays a critical role in preserving dermal elasticity and preventing extracellular matrix degradation. Thus, employing ultrasound-assisted green extraction under optimized conditions represents a sustainable and efficient strategy for maximizing the recovery of compounds with anti-inflammatory, antioxidant, epithelialization-promoting, and elastase-inhibitory properties. From a broader perspective, this integrative approach underscores how eco-friendly extraction technologies can bridge natural product chemistry and functional efficacy, paving the way for the development of scientifically grounded, sustainable anti-aging formulations that combine biological potency with environmental responsibility.

3. Materials and Methods

3.1. Raw Materials

The fruits of *J. regia* with green husk were harvested from Çamalanı, Akkuş, Ordu, Türkiye (40°49′41″ N 37°00′54″ E) (AEF26880). The green husks were carefully removed from the walnuts and subsequently dried to a stable weight in an incubator set at 35 °C for approximately 3 days. Following this process, the plant material was ground to approximately 40 mesh (0.42 mm), and the resulting homogeneous sample was stored in a desiccator, shielded from light, for future analyses [71].

3.2. Preparation of NADES

In the preliminary phase of this study, several NADES combinations were systematically prepared and evaluated at both 1:1 and 1:2 molar ratios between the HBA and HBD components. During this selection process, particular emphasis was placed on using naturally derived and biodegradable components that align with green chemistry principles. The formulations tested included ChCl:glucose (1:1, 1:2), ChCl:sucrose (1:1, 1:2), ChCl:lactic acid (1:1, 1:2), citric acid:sucrose (1:1, 1:2), and citric acid:glucose (1:1, 1:2) [72,73]. Among these, the ChCl:glycerol (1:2) system displayed the most favorable balance of physicochemical properties, superior phase homogeneity, and remarkable stability under UAE conditions. While certain sugar- and acid-based NADESs demonstrated inadequate solubilization potential, others with relatively acceptable solubility still suffered from elevated viscosity and poor thermal stability, factors that likely restricted effective cavitation and compromised overall extraction efficiency. In contrast, the ChCl:glycerol (1:2) mixture maintained a clear, low-viscosity phase and demonstrated structural stability during sonication, indicating its suitability as a thermally robust solvent. In particular, this system exhibited thermal stability within the operational temperature range applied during UAE, supporting its selection for subsequent optimization.

Consistent with our experimental findings, the ChCl:glycerol (1:2) NADES was selected due to its thermal stability, low viscosity, and biocompatibility compared with other tested systems. Its flexible hydrogen-bond network maintains structural integrity under ultrasonic and thermal conditions, ensuring efficient mass transfer and stable extraction performance [47,72]. The system also benefits from the GRAS status of components, providing a safe, low-cytotoxic, and highly biodegradable extraction medium suitable for applications in food, cosmetic, and pharmaceutical formulations [44,47,74]. Overall, these findings—supported by both preliminary solvent screening and literature evidence—confirmed the selection of the ChCl:glycerol (1:2) NADES as the optimal and most sustainable solvent system for the present study [43,75].

NADES was synthesized following the methodology described by Sakti et al. [18]. The preparation involved combining ChCl (CAS Number: 67-48-1, Cayman Chemical, ≥95%) as the HBA with glycerol (CAS Number: 56-81-5, Merck, ≥99.5%) as the HBD in a 1:2 (*w*/*w*) ratio. The mixture was stirred in a glass beaker using a magnetic stirrer (IKA® C-MAG HS 7, IKA-Werke GmbH & Co. KG, Staufen, Germany) at 80 °C. The beaker was then sealed with Parafilm and left to rest for approximately one hour, or until a clear liquid was obtained, after which it was used immediately.

3.3. Preparations of Extracts with NADES

The extraction process utilizing NADES was conducted with an ultrasonicator (ISOLAB Ultrasonic C., Catalogue No:621.05.003, Eschau, Germany) operating at a power of 35 W and a frequency of 42.000 Hz. Powdered samples were placed in vials and combined with deionized water (dH_2O) as outlined in Table 4. Subsequently, NADES was introduced into the vials. The UAE was carried out with different ratios of NADES-H_2O mixture (25%, 62.5%, 100%), different temperatures (40 °C, 60 °C, 80 °C), and different times (20 min, 40 min, 60 min) (Table 4). Following this, the mixtures were transferred to centrifuge tubes and centrifuged for 10 min at a force of 3.283 g to facilitate the separation of the NADES liquid extract from the residual material. The resulting liquid layer was then filtered through a 0.45 µm Whatman micropore filter paper [18]. Subsequently, all extracts were subjected to evaporation under reduced pressure at 40 °C to yield crude extracts.

Table 4. Independent and dependent variables of BBD Analysis.

		Independent Variables			
Factors	**Name**	**Units**	**Coded Low**	**Coded Mean**	**Coded High**
A	NADES/H_2O ratio *	%	−1 ↔ 25.00	0 ↔ 62.50	+1 ↔ 100.00
B	Temperature	°C	−1 ↔ 40.00	0 ↔ 60.00	+1 ↔ 80.00
C	Time	min.	−1 ↔ 20.00	0 ↔ 40.00	+1 ↔ 60.00
		Dependent Variables			
Responses	Name	Units			
R1	Elastaz inhibition (IC_{50})	µg/mL			

* NADES/H_2O (*w*/*w*).

After the extraction process, the NADES phase was removed and treated as waste, since trace amounts of minor components could potentially remain and interfere with subsequent analyses. Nevertheless, due to its biodegradable nature and low environmental impact, the discarded solvent can still be regarded as an environmentally friendly system.

3.4. Evaluation of Elastase Inhibition

The elastase inhibitory activity was determined spectrophotometrically according to the method described by Moon et al. [76] and Sun et al. [66], with minor modifications. A volume of 50 µL of each extract solution, prepared at varying concentrations in 0.2 M Tris-HCl buffer (pH 7.8) containing 0.01% (*v*/*v*) dimethyl sulfoxide (DMSO), was mixed with 100 µL of HNE solution (17 mU/mL) prepared in the same buffer. For the control, 100 µL of HNE solution and 50 µL of buffer containing 0.01% DMSO were used without the addition of the extract. The blank contained 100 µL of enzyme-free phosphate buffer and 50 µL of phosphate buffer containing 0.01% DMSO. All tubes (sample, control, and blank) were incubated at 37 °C for 15 min. Subsequently, 50 µL of 5 mM N-succinyl-Ala-Ala-Ala-*p*-nitroanilide (STANA) substrate solution was added, and the mixtures were further incubated at 37 °C for 30 min. The absorbance values of the control and sample solutions were measured at 410 nm against the blank using a spectrophotometer. Catechin was used as the standard. The tests were conducted in triplicate, and necessary calculations were performed based on the average absorbance values. The elastase inhibition activity of the prepared plant extracts and the standard substance was calculated using the following Formula (2):

$$\% \text{ inhibition} = [(A_{control} - A_{sample})/A_{control}] \times 100 \quad (2)$$

where $A_{control}$ represents the absorbance value of the control solution at 410 nm, and A_{sample} denotes the absorbance value of the sample solution at the same wavelength. The IC_{50} value of the elastase enzyme was determined from the regression equation derived from the linear fit of the plotted data of concentration versus % elastase inhibition.

3.5. *Optimization of Extraction Using Response Surface Methodology (RSM)*

The optimization study was conducted utilizing a factorial-based BBD (3-factor, 3-level) as implemented in Design Expert V.8 [77]. This investigation focused on a dependent variable: elastase inhibition, while three independent variables: solvent ratio, temperature, and time were systematically analyzed. The selection of these factors and their corresponding levels was informed by prior research [78–82]. The coded values and their associated levels are presented in Table 4. The experimental design comprised a total of 17 runs, including five center points. The extraction conditions for all 17 runs are detailed in Table 1.

Also, to compare the extraction technique, the maceration process was similarly utilized with the same NADES/H_2O ratios under optimal conditions at ambient temperature for one day, allowing for a comparison with both the green extraction method and the conventional technique.

3.6. *Statistical Analysis*

The tests were conducted in triplicate, and the results are expressed as mean ± standard deviation. IC_{50} values were determined from the regression equation of a linear curve plotted by the logarithm of the concentrations against the percentage inhibition values. Statistical analyses were performed using IBM SPSS Statistics software (Version 24.0) [83]. Principal Component Analysis (PCA) was performed to visualize the overall variance and identify clustering patterns among samples obtained from different extraction methods (U1, U2, and M1). PCA is an unsupervised multivariate statistical technique that reduces the dimensionality of complex datasets while retaining most of the variation present in the data [24]. In this study, PCA was applied to assess the reproducibility and discrimination capability of the extraction protocols. The first two principal components explained the majority of total variance, allowing the evaluation of similarities and differences in metabolite profiles across extraction methods.

3.7. *Metabolomics Analysis*

Metabolomics analyses were performed on the extracts derived from M1, U1, and U2. U1 was characterized as the extract obtained through ultrasonic extraction under optimal conditions, specifically aimed at elastase inhibition. While U2 refers to the extract prepared under optimal conditions with the same solvent ratio using the ultrasonic extraction technique, but at room temperature, M1 represents the extract prepared under optimal conditions with the same solvent ratio by the maceration technique for 24 h at room temperature.

Metabolite separation was performed using a Zorbax C18 column (1 × 50 mm, 1.8 µm, 100 Å) and analyzed through liquid chromatography–quadrupole time-of-flight mass spectrometry (LC-qTOF-MS) system (Agilent 6530). The mobile phase included solvent A (water containing 0.1% formic acid) and solvent B (acetonitrile with 0.1% formic acid), following a gradient elution method: 0–1 min, 10% B; 1–10 min, 10–90% B; 10–11 min, 90%

B; 11–20 min, 90–10% B; 20–30 min, 10% B. The flow rate was set at 0.2 mL/min, with a 2 μL injection volume. The ESI source functioned in positive mode, with a capillary voltage of 4000 V and a capillary temperature of 300 °C. Auto MS-MS data acquisition covered an m/z range of 100 to 1700, with a minimum threshold of 200 counts. MS/MS fragmentation of plant metabolites was carried out at a collision energy of 20 eV.

Raw MS data were analyzed using MS-Dial version 2.56 for deconvolution, peak detection, and alignment [24]. The minimum peak height for detection was set at 2000 amplitude, with MS1 and MS2 tolerances of 0.01 and 0.025 Da, respectively. Molecular formula prediction and structural elucidation were performed using MS-FINDER version 3.04 [84]. Identifications were refined through MS/MS data analysis. Metabolites were matched against natural product databases, including the Universal Natural Products Database (UNPD), KNAPpSAcK, and PlantCyc, with a mass tolerance of 10 ppm. Only metabolites with a score higher than 6 were regarded as confidently identified.

4. Conclusions

This research aimed to determine the optimal extraction conditions for the green husk of walnuts by targeting elastase inhibition through BBD analysis. During the extraction process, environmentally friendly NADESs were employed. As a result of the study, the walnut green husk extract prepared through eco-friendly methods demonstrated a high elastase inhibition effect, revealing its anti-aging potential. Metabolomic analysis indicated an increase in phenolic compound content under the optimized conditions. The enhanced anti-aging potential was associated with an increase in both the number and rate of phenolic compounds. These findings emphasize the feasibility of implementing environmentally friendly and sustainable approaches through the valorization of natural waste, green extraction techniques, and process optimization. However, further research and clinical studies are required to support the efficacy of the green husk of walnuts in this regard.

Supplementary Materials: The following supporting information can be downloaded at: https://www.mdpi.com/article/10.3390/molecules30214191/s1, Table S1. Results of the BBD's model adequacy test. Table S2. List of possible identified metabolites of UMS using ESI–MS/MS in the negative ionization mode ($[M-H]^-$) and advanced correlation analysis data. Figure S1. Diagnostic plots for the Box–Behnken model accuracy of predicted and actual data against elastase inhibition. Predicted versus actual values for elastase inhibition in the Box–Behnken model. The color scale indicates the magnitude of the response, where blue represents lower inhibition levels and red represents higher inhibition levels. Figure S2. Hierarchical clustering heat map showing the differential distribution of metabolites among the extracts. The color scale represents normalized relative abundance values ranging from 1 (lowest abundance, blue) to +1 (highest abundance, red).

Author Contributions: Conceptualization, S.Ö.Ş. and S.S.; methodology, S.Ö.Ş., S.S.B., Ö.Ş. and E.K.; software, S.Ö.Ş., Ö.Ş. and E.K.; validation, S.Ö.Ş. and E.K.; formal analysis, S.Ö.Ş. and E.K.; investigation, S.Ö.Ş., S.S. and S.S.B.; resources, S.S.B. and Ş.N.K.; data curation, S.Ö.Ş., S.S.B. and Ö.Ş.; writing—original draft preparation, S.Ö.Ş., S.S. and T.S.; writing—review and editing S.Ö.Ş., S.S. and T.S.; visualization, S.Ö.Ş., S.S.B. and Ö.Ş.; supervision, S.Ö.Ş., E.K. and E.N.; project administration, S.Ö.Ş. and S.S.B.; funding acquisition, S.Ö.Ş. and S.S. All authors have read and agreed to the published version of the manuscript.

Funding: This study was funded by The Scientific and Technological Research Council of Türkiye (TUBITAK), grant number: 1919B012303375.

Institutional Review Board Statement: Not applicable.

Informed Consent Statement: Not applicable.

Data Availability Statement: Data are contained within the article.

Acknowledgments: The authors gratefully acknowledge Merve Badem for her valuable contribution to the identification of the plant material.

Conflicts of Interest: The authors declare no conflicts of interest.

Abbreviations

The following abbreviations are used in this manuscript:

WHO	World Health Organization
USA	United States of America
UV	Ultraviolet
ILs	Ionic liquids
DES	Deep eutectic solvents
NADES	Natural Deep Eutectic Solvents
HBD	Hydrogen bond donor
BBD	Box–Behnken design
UAE	Ultrasound-assisted extraction
IC_{50}	Half maximal inhibitory concentration
U1	The BBD-optimized extract
M1	The Maceration extract
U2	The Optimized extract at maceration temperature
2FI	Two-factor interaction
R^2	Coefficient of determination
PRE	Percent relative error
ANOVA	Analysis of variance
CV%	Coefficient of variance
R^2a	Adjusted coefficient of determination
GAE	Gallic acid equivalent
dw	Dry weight
RE	Rutin equivalent
ChCl	Choline chloride
RSM	Response surface methodology
PCA	Principal Component Analysis
ROS	Reactive oxygen species
ERK1/2	Extracellular signal-regulated kinase 1/2
Nrf2	Nuclear factor erythroid 2-related factor 2
NO	Nitric oxide
HNE	Human neutrophil elastase
PGE2	Prostaglandin E2
IL	Interleukin
Dct	Dopachrome tatamerase
TRP-1	Tyrosinase-related protein-1
MITF	Microphthalmia-associated transcription factor
cAMP	Cyclic adenosine monophosphate
CREB	cAMP-responsive binding protein
dH_2O	Deionized water
HBA	Hydrogen bond acceptor
DMSO	Dimethyl sulfoxide
STANA	N-succinyl-ala-ala-ala-*p*-nitroanilide
LC-qTOF-MS	Liquid chromatography–quadrupole time-of-flight mass spectrometry
UNPD	Universal Natural Products Database

References

1. World Health Organization. Ageing: Healthy Ageing and Functional Ability. 2020. Available online: https://www.who.int/news-room/questions-and-answers/item/healthy-ageing-and-functional-ability (accessed on 21 October 2025).
2. Ganceviciene, R.; Liakou, A.I.; Theodoridis, A.; Makrantonaki, E.; Zouboulis, C.C. Skin anti-aging strategies. *Dermato-Endocrinology* **2012**, *4*, 308–319. [CrossRef]
3. Öztürkcan, S.; Kayhan, T.Ç. Deri yaşlanmasına karşı medikal önlemler. *Dermatoz* **2010**, *1*, 29–34.
4. Global Industry Analysts. Anti-Aging Products-Global Market Trajectory & Analytics. Dublin. 2020. Available online: https://www.researchandmarkets.com/r/jnrvp0 (accessed on 23 June 2022).
5. Vijayakumar, R.; Gani, S.S.A.; Mokhtar, N. Anti-elastase, anti-collagenase and antimicrobial activities of the underutilized red pitaya peel: An in vitro study for anti-aging applications. *Asian J. Pharm. Clin. Res.* **2017**, *10*, 251–255. [CrossRef]
6. Imokawa, G.; Ishida, K. Biological mechanisms underlying the ultraviolet radiation-induced formation of skin wrinkling and sagging I: Reduced skin elasticity, highly associated with enhanced dermal elastase activity, triggers wrinkling and sagging. *Int. J. Mol. Sci.* **2015**, *16*, 7753–7775. [CrossRef]
7. Tsuji, N.; Moriwaki, S.; Suzuki, Y.; Takema, Y.; Imokawa, G. The role of elastases secreted by fibroblasts in wrinkle formation: Implication through selective inhibition of elastase activity. *Photochem. Photobiol.* **2001**, *74*, 283–290. [CrossRef]
8. Jahanban-Esfahlan, A.; Amarowicz, R. Walnut (*Juglans regia* L.) shell pyroligneous acid: Chemical constituents and functional applications. *RSC Adv.* **2018**, *8*, 22376–22391. [CrossRef]
9. Przekora, A.; Belcarz, A.; Kowalczyk, K.; Wójcik, M.; Wojciechowska, K.; Ginalska, G. UVB protective, anti-aging, and anti-inflammatory properties of aqueous extract of walnut (*Juglans regia* L.) seeds. *Acta Pol. Pharm.* **2018**, *75*, 1167–1176. [CrossRef]
10. Li, B.; Cui, C.; Zhang, C.; Liu, J.; Hao, F.; Han, L.; Bai, C.; Wei, S. Traditional applications, ethnopharmacology, and phytochemistry of walnut green husk (*Juglans regia* L.): A review. *Nat. Prod. Commun.* **2024**, *19*, 1934578X241262156. [CrossRef]
11. Jahanban-Esfahlan, A.; Ostadrahimi, A.; Tabibiazar, M.; Amarowicz, R. A comprehensive review on the chemical constituents and functional uses of walnut (*Juglans* spp.) husk. *Int. J. Mol. Sci.* **2019**, *20*, 3920. [CrossRef]
12. Liu, M.; Li, C.; Cao, C.; Wang, L.; Li, X.; Che, J.; Yang, H.; Zhang, X.; Zhao, H.; He, G.; et al. Walnut fruit processing equipment: Academic insights and perspectives. *Food Eng. Rev.* **2021**, *13*, 822–857. [CrossRef]
13. Barekat, S.; Nasirpour, A.; Keramat, J.; Dinari, M.; Meziane-Kaci, M.; Paris, C.; Desobry, S. Phytochemical composition, antimicrobial, anticancer properties, and antioxidant potential of green husk from several walnut varieties (*Juglans regia* L.). *Antioxidants.* **2023**, *12*, 52. [CrossRef] [PubMed]
14. Chemat, F.; Abert-Vian, M.; Fabiano-Tixier, A.S.; Strube, J.; Uhlenbrock, L.; Gunjevic, V.; Cravotto, G. Green extraction of natural products. Origins, current status, and future challenges. *Trend. Anal. Chem.* **2019**, *118*, 248–263. [CrossRef]
15. Demirdoven, A.D.; Tokatli, K.; Korkmaz, Y. Geleneksel ve ultrasonik yöntemlerle vişne posası antosiyaninlerinin ekstraksiyonu. *Gıda* **2020**, *46*, 168–179. [CrossRef]
16. Çağındı, Ö. Mikrodalga uygulamasının kırmızı üzüm suyunun antosiyanin içeriği ile bazı fizikokimyasal özellikleri üzerine etkisi. *Akademik Gıda* **2016**, *14*, 356–361.
17. Picot-Allain, C.; Mahomoodally, M.F.; Ak, G.; Zengin, G. Conventional versus green extraction techniques—A comparative perspective. *Curr. Opin. Food Sci.* **2021**, *40*, 144–156. [CrossRef]
18. Sakti, A.S.; Saputri, F.C.; Mun'im, A. Optimization of choline chloride-glycerol based natural deep eutectic solvent for extraction bioactive substances from *Cinnamomum burmannii* barks and *Caesalpinia sappan* heartwoods. *Heliyon* **2019**, *5*, e02915. [CrossRef]
19. Kutlu, N.; Kamiloglu, A.; Elbir, T. Optimization of ultrasound extraction of phenolic compounds from Tarragon (*Artemisia dracunculus* L.) using Box–Behnken design. *Biomass Conv. Bioref.* **2022**, *12*, 5397–5408. [CrossRef]
20. Ferreira, S.C.; Bruns, R.E.; Ferreira, H.S.; Matos, G.D.; David, J.M.; Brandão, G.C.; Souza, A.S.; Dos Santos, W.N.L. Box-Behnken design: An alternative for the optimization of analytical methods. *Anal. Chim. Acta* **2007**, *597*, 179–186. [CrossRef]
21. Khajeh, M. Optimization of process variables for essential oil components from *Satureja hortensis* by supercritical fluid extraction using Box-Behnken experimental design. *J. Supercrit. Fluid.* **2011**, *55*, 944–948. [CrossRef]
22. Eshawu, A.B.; Ghalsasi, V.V. Metabolomics of natural samples: A tutorial review on the latest technologies. *J. Sep. Sci.* **2024**, *47*, 2300588. [CrossRef] [PubMed]
23. Chitiva, L.C.; Rezende-Teixeira, P.; Leão, T.F.; Lozano-Puentes, H.S.; Londoño, X.; Díaz-Ariza, L.A.; Costa-Lotufo, L.V.; Prieto-Rodríguez, J.A.; Costa, G.M.; Castro-Gamboa, I. Metabolomic profiling of *Guadua* species and its correlation with antioxidant and cytotoxic activities. *ACS Omega* **2024**, *9*, 36939–36960. [CrossRef]
24. Gonulalan, E.M.; Nemutlu, E.; Bayazeid, O.; Kocak, E.; Yalçın, F.N.; Demirezer, L.O. Metabolomics and proteomics profiles of some medicinal plants and correlation with BDNF activity. *Phytomedicine* **2020**, *74*, 152920. [CrossRef] [PubMed]
25. Csekes, E.; Račková, L. Skin aging, cellular senescence and natural polyphenols. *Int. J. Mol. Sci.* **2021**, *22*, 12641. [CrossRef] [PubMed]
26. Tan, M.C.; Chin, N.L.; Yusof, Y.A. A Box–Behnken design for determining the optimum experimental condition of cake batter mixing. *Food Bioprocess Technol.* **2012**, *5*, 972–982. [CrossRef]

27. Sankalia, M.G.; Mashru, R.C.; Sankalia, J.M.; Sutariya, V.B. Reversed chitosan–alginate polyelectrolyte complex for stability improvement of alpha-amylase: Optimization and physicochemical characterization. *Eur. J. Pharm. Biopharm.* **2007**, *65*, 215–232. [CrossRef]
28. Maran, J.P.; Mekala, V.; Manikandan, S. Modeling and optimization of ultrasound-assisted extraction of polysaccharide from *Cucurbita moschata*. *Carbohydr. Polym.* **2013**, *92*, 2018–2026. [CrossRef]
29. Shakeel, F.; Haq, N.; Raish, M.; Anwer, M.K.; Al-Shdefat, R. Solubility and thermodynamic analysis of sinapic acid in various neat solvents at different temperatures. *J. Mol. Liq.* **2016**, *222*, 167–171. [CrossRef]
30. Shen, L.; Pang, S.; Zhong, M.; Sun, Y.; Qayum, A.; Liu, Y.; Rashid, A.; Xu, B.; Liang, Q.; Ma, H.; et al. A comprehensive review of ultrasonic assisted extraction (UAE) for bioactive components: Principles, advantages, equipment, and combined technologies. *Ultrason. Sonochem.* **2023**, *101*, 106646. [CrossRef] [PubMed]
31. Chen, C.Y.; Wang, G.H.; Kuo, J.T.; Lin, Y.T.; Huang, H.J.; Chang, Y.C.; Chung, Y.C. Ultrasound-assisted extraction of *Peucedanum ostruthium* leaves: A feasible alternative to rhizomes for industrial applications. *Front. Pharmacol.* **2025**, *16*, 1636312. [CrossRef]
32. Bimakr, M.; Rahman, R.A.; Taip, F.S.; Adzahan, N.M.; Sarker, M.Z.I.; Ganjloo, A. Optimization of ultrasound-assisted extraction of crude oil from winter melon (*Benincasa hispida*) seed using response surface methodology and evaluation of its antioxidant activity, total phenolic content and fatty acid composition. *Molecules* **2012**, *17*, 11748–11762. [CrossRef]
33. Kumar, K.; Srivastav, S.; Sharanagat, V.S. Ultrasound assisted extraction (UAE) of bioactive compounds from fruit and vegetable processing by-products: A review. *Ultrason. Sonochem.* **2021**, *70*, 105325. [CrossRef]
34. Cui, L.; Ma, Z.; Wang, D.; Niu, Y. Ultrasound-assisted extraction, optimization, isolation, and antioxidant activity analysis of flavonoids from *Astragalus membranaceus* stems and leaves. *Ultrason. Sonochem.* **2022**, *90*, 106190. [CrossRef] [PubMed]
35. Hu, D.B.; Xue, R.; Zhuang, X.C.; Zhang, X.S.; Shi, S.L. Ultrasound-assisted extraction optimization of polyphenols from *Boletus bicolor* and evaluation of its antioxidant activity. *Front. Nutr.* **2023**, *10*, 1135712. [CrossRef]
36. Slimani, C.; Rais, C.; Mansouri, F.; Rais, S.; Benjelloun, M.; Ullah, R.; Lazraq, A. Optimization of ultrasound-assisted extraction of phenols from *Crocus sativus* by-products using sunflower oil as a sustainable solvent alternative. *Food Chem. X* **2024**, *23*, 101579. [CrossRef]
37. Jiang, C.; Li, X.; Jiao, Y.; Jiang, D.; Zhang, L.; Fan, B.; Zhang, Q. Optimization for ultrasound-assisted extraction of polysaccharides with antioxidant activity in vitro from the aerial root of *Ficus microcarpa*. *Carbohydr. Polym.* **2014**, *110*, 10–17. [CrossRef]
38. Thao My, P.L.; Sung, V.V.; Dat, T.D.; Nam, H.M.; Phong, M.T.; Hieu, N.H. Ultrasound-assisted extraction of fucoidan from Vietnamese brown seaweed *Sargassum mcclurei* and testing bioactivities of the extract. *ChemistrySelect* **2020**, *5*, 4371–4380. [CrossRef]
39. Drouet, S.; Leclerc, E.A.; Garros, L.; Tungmunnithum, D.; Kabra, A.; Abbasi, B.H.; Hano, C. A green ultrasound-assisted extraction optimization of the natural antioxidant and anti-aging flavonolignans from milk thistle *Silybum marianum* (L.) Gaertn. fruits for cosmetic applications. *Antioxidants* **2019**, *8*, 304. [CrossRef]
40. Landa-Cansigno, C.; Serviere-Zaragoza, E.; Reyes, A.G.; Monribot-Villanueva, J.L.; Guerrero-Analco, J.A.; Sandoval-Gil, J.M.; Zertuche-González, J.A. Untargeted metabolomics-driven insights into the anti-aging potential of brown seaweed *Eisenia arborea*: Ultrasound-assisted extraction for sustainable cosmetics. *J. Appl. Phycol.* **2025**, *37*, 1–13. [CrossRef]
41. Alshammari, F.; Alam, M.B.; Naznin, M.; Kim, S.; Lee, S.H. Optimization, metabolomic analysis, antioxidant potential and depigmenting activity of polyphenolic compounds from unmature ajwa date seeds (*Phoenix dactylifera* L.) using ultrasonic-assisted extraction. *Antioxidants* **2024**, *13*, 238. [CrossRef] [PubMed]
42. Gouvinhas, I.; Saavedra, M.J.; Alves, M.J.; Garcia, J. Exploring the impact of ultrasound-assisted extraction on the phytochemical composition and bioactivity of *Tamus communis* L. fruits. *Pharmaceuticals* **2025**, *18*, 1342. [CrossRef] [PubMed]
43. Dai, Y.; Van Spronsen, J.; Witkamp, G.J.; Verpoorte, R.; Choi, Y.H. Natural deep eutectic solvents as new potential media for green technology. *Anal. Chim. Acta* **2013**, *766*, 61–68. [CrossRef]
44. Paiva, A.; Craveiro, R.; Aroso, I.; Martins, M.; Reis, R.L.; Duarte, A.R.C. Natural deep eutectic solvents—Solvents for the 21st century. *ACS Sustain. Chem. Eng.* **2014**, *2*, 1063–1071. [CrossRef]
45. Chanioti, S.; Tzia, C. Extraction of phenolic compounds from olive pomace by using natural deep eutectic solvents and innovative extraction techniques. *Innov. Food Sci. Emerg. Technol.* **2018**, *48*, 228–239. [CrossRef]
46. Pavlić, B.; Mrkonjić, Ž.; Teslić, N.; Kljakić, A.C.; Pojić, M.; Mandić, A.; Mišan, A. Natural deep eutectic solvent (NADES) extraction improves polyphenol yield and antioxidant activity of wild thyme (*Thymus serpyllum* L.) extracts. *Molecules* **2022**, *27*, 1508. [CrossRef]
47. Ferreira, A.S.; Craveiro, R.; Duarte, A.R.; Barreiros, S.; Cabrita, E.J.; Paiva, A. Effect of water on the structure and dynamics of choline chloride/glycerol eutectic systems. *J. Mol. Liq.* **2021**, *342*, 117463. [CrossRef]
48. Dai, Y.; Witkamp, G.J.; Verpoorte, R.; Choi, Y.H. Tailoring properties of natural deep eutectic solvents with water to facilitate their applications. *Food Chem.* **2015**, *187*, 14–19. [CrossRef] [PubMed]
49. Bildik, F. Application of choline chloride as natural deep eutectic solvents for the green extraction of phenolic compounds from *Rheum ribes* leaves. *Eur. J. Tech.* **2021**, *11*, 140–143. [CrossRef]

50. Frontini, A.; Tarentini, G.; Negro, C.; Luvisi, A.; Apollonio, M.; De Bellis, L. Effective liquid–liquid extraction for the recovery of grape pomace polyphenols from natural deep eutectic solvents (NaDES). *Separations* **2025**, *12*, 148. [CrossRef]
51. García-Roldán, A.; Piriou, L.; Jauregi, P. Natural deep eutectic solvents as a green extraction of polyphenols from spent coffee ground with enhanced bioactivities. *Front. Plant Sci.* **2023**, *13*, 1072592. [CrossRef] [PubMed]
52. Swamy, G.J.; Sangamithra, A.; Chandrasekar, V. Response surface modeling and process optimization of aqueous extraction of natural pigments from *Beta vulgaris* using Box–Behnken design of experiments. *Dyes Pigm.* **2014**, *111*, 64–74. [CrossRef]
53. Stoica, F.; Constantin, O.E.; Stănciuc, N.; Aprodu, I.; Bahrim, G.E.; Râpeanu, G. Optimization of the parameters influencing the antioxidant activity and concentration of anthocyanins extracted from red onion skins using a central composite design. *Inventions* **2022**, *7*, 89. [CrossRef]
54. Radhakrishnan, N.; Prabhakaran, V.S.; Wadaan, M.A.; Baabbad, A.; Vinayagam, R.; Kang, S.G. STITCH, physicochemical, ADMET, and in silico analysis of selected *Mikania* constituents as anti-inflammatory agents. *Processes* **2023**, *11*, 1722. [CrossRef]
55. Deniz, F.S.S.; Salmas, R.E.; Emerce, E.; Cankaya, I.I.T.; Yusufoglu, H.S.; Orhan, I.E. Evaluation of collagenase, elastase and tyrosinase inhibitory activities of *Cotinus coggygria* Scop. through in vitro and in silico approaches. *S. Afr. J. Bot.* **2020**, *132*, 277–288. [CrossRef]
56. Wittenauer, J.; Mäckle, S.; Sußmann, D.; Schweiggert-Weisz, U.; Carle, R. Inhibitory effects of polyphenols from grape pomace extract on collagenase and elastase activity. *Fitoterapia* **2015**, *101*, 179–187. [CrossRef]
57. Wu, Y.; Ding, C.; Zhang, Z.; Zhang, J.; Li, Y.; Song, X.; Zhang, D. Sesquilignans: Current research and potential prospective. *Eur. J. Med. Chem.* **2024**, *271*, 116445. [CrossRef] [PubMed]
58. Aydın, T.; Gümüştaş, M.; Sancı, T.Ö.; Çakır, A. Herniarin and skimmin coumarins in spice and edible plants and their benefits for health. In *Studies in Natural Products Chemistry*; Atta-Ur-Rahman, Ed.; Elsevier: Amsterdam, The Netherlands, 2024; Volume 81, pp. 339–365. [CrossRef]
59. Chang, K.C.; Laffin, B.; Ponder, J.; Énzsöly, A.; Németh, J.; LaBarbera, D.V.; Petrash, J.M. Beta-glucogallin reduces the expression of lipopolysaccharide-induced inflammatory markers by inhibition of aldose reductase in murine macrophages and ocular tissues. *Chem. Biol. Interact.* **2013**, *202*, 283–287. [CrossRef]
60. Kim, H.G.; Kim, K.S.; Kim, M.; Shin, S.H.; Lee, Y.G.; Bang, M.H.; Lee, D.G.; Baek, N.I. β-Glucogallin isolated from *Fusidium coccineum* and its enhancement of skin barrier effects. *Appl. Biol. Chem.* **2020**, *63*, 77. [CrossRef]
61. Liang, H.; Huang, Q.; Zou, L.; Wei, P.; Lu, J.; Zhang, Y. Methyl gallate: Review of pharmacological activity. *Pharmacol. Res.* **2023**, *194*, 106849. [CrossRef] [PubMed]
62. Kumar, K.S.; Vani, M.G.; Wang, S.Y.; Liao, J.W.; Hsu, L.S.; Yang, H.L.; Hseu, Y.C. In vitro and in vivo studies disclosed the depigmenting effects of gallic acid: A novel skin lightening agent for hyperpigmentary skin diseases. *Biofactors* **2013**, *39*, 259–270. [CrossRef]
63. Lee, I.K.; Yun, B.S. Styrylpyrone-class compounds from medicinal fungi *Phellinus* and *Inonotus* spp., and their medicinal importance. *J. Antibiot.* **2011**, *64*, 349–359. [CrossRef]
64. Kim, M.; Kim, J.; Shin, Y.K.; Kim, K.Y. Gentisic acid stimulates keratinocyte proliferation through ERK1/2 phosphorylation. *Int. J. Med. Sci.* **2020**, *17*, 626. [CrossRef]
65. Ma, N.S.; Abraham, M.; Choi, M.J.; Kim, J.S. Skin permeation and comparative evaluation of gentisic acid ester derivatives as skin-lightening agents. *J. Drug Deliv. Sci. Technol.* **2014**, *24*, 212–217. [CrossRef]
66. Sun, Y.; Zhang, M.; Zhang, Y.; Zheng, Y.; Li, J.; Cai, Q.; Qu, Y. Synergistic and antagonistic mechanisms of *Arctium lappa* L. polyphenols on human neutrophil elastase inhibition: Insights from molecular docking and enzymatic kinetics. *Molecules* **2025**, *30*, 2764. [CrossRef]
67. Tóth, B.; Chang, F.R.; Hwang, T.L.; Szappanos, Á.; Mándi, A.; Hunyadi, A.; Vasas, A. Screening of *Luzula* species native to the Carpathian Basin for anti-inflammatory activity and bioactivity-guided isolation of compounds from *Luzula luzuloides* (Lam.) Dandy & Wilmott. *Fitoterapia* **2017**, *116*, 131–138. [CrossRef]
68. Girsang, E.; Lister, I.N.E.; Ginting, C.N.; Bethasari, M.; Amalia, A.; Widowati, W. Comparison of antiaging and antioxidant activities of protocatechuic and ferulic acids. *Mol. Cell. Biomed. Sci.* **2020**, *4*, 68–75. [CrossRef]
69. Xing, X.; Yang, X.; Cao, Y. Study of ellagic acid as a natural elastase inhibitor by spectroscopic methods. *J. Appl. Spectrosc.* **2016**, *83*, 149–155. [CrossRef]
70. Maity, N.; Nema, N.K.; Abedy, M.K.; Sarkar, B.K.; Mukherjee, P.K. Exploring *Tagetes erecta* Linn flower for the elastase, hyaluronidase and MMP-1 inhibitory activity. *J. Ethnopharmacol.* **2011**, *137*, 1300–1305. [CrossRef] [PubMed]
71. Vieira, V.; Pereira, C.; Abreu, R.M.; Calhelha, R.C.; Alves, M.J.; Coutinho, J.A.P.; Ferreira, O.; Barros, L.; Ferreira, I.C. Hydroethanolic extract of *Juglans regia* L. green husks: A source of bioactive phytochemicals. *Food Chem. Toxicol.* **2020**, *137*, 111189. [CrossRef] [PubMed]
72. Skarpalezos, D.; Detsi, A. Deep eutectic solvents as extraction media for valuable flavonoids from natural sources. *Appl. Sci.* **2019**, *9*, 4169. [CrossRef]

73. Rolińska, K.; Jakubowska, E.; Żmieńko, M.; Łęczycka-Wilk, K. Choline chloride-based deep eutectic solvents as plasticizer and active agent in chitosan films. *Food Chem.* **2024**, *444*, 138375. [CrossRef]
74. Huang, M.M.; Yiin, C.L.; Lock, S.S.M.; Chin, B.L.F.; Othman, I.; Chan, Y.H. Natural deep eutectic solvents (NADES) for sustainable extraction of bioactive compounds from medicinal plants: Recent advances, challenges, and future directions. *J. Mol. Liq.* **2025**, *425*, 127202. [CrossRef]
75. Freitas, D.S.; Rocha, D.; Castro, T.G.; Noro, J.; Castro, V.I.; Teixeira, M.A.; Silva, C. Green extraction of cork bioactive compounds using natural deep eutectic mixtures. *ACS Sustain. Chem. Eng.* **2022**, *10*, 7974–7989. [CrossRef]
76. Moon, J.Y.; Yim, E.Y.; Song, G.; Lee, N.H.; Hyun, C.G. Screening of elastase and tyrosinase inhibitory activity from Jeju Island plants. *Eurasia J. Biosci.* **2010**, *4*, 41–53. [CrossRef]
77. Abd-El-Aziz, N.M.; Hifnawy, M.S.; El-Ashmawy, A.A.; Lotfy, R.A.; Younis, I.Y. Application of Box–Behnken design for optimization of phenolics extraction from *Leontodon hispidulus* in relation to its antioxidant, anti-inflammatory and cytotoxic activities. *Sci. Rep.* **2022**, *12*, 8829. [CrossRef]
78. Ahmad, A.; Rehman, M.U.; Wali, A.F.; El-Serehy, H.A.; Al-Misned, F.A.; Maodaa, S.N.; Aljawdah, H.M.; Mir, T.M.; Ahmad, P. Box–Behnken response surface design of polysaccharide extraction from *Rhododendron arboreum* and the evaluation of its antioxidant potential. *Molecules* **2020**, *25*, 3835. [CrossRef] [PubMed]
79. Kaiser, S.; Verza, S.G.; Moraes, R.C.; Pittol, V.; Peñaloza, E.M.C.; Pavei, C.; Ortega, G.G. Extraction optimization of polyphenols, oxindole alkaloids and quinovic acid glycosides from cat's claw bark by Box–Behnken design. *Ind. Crops Prod.* **2013**, *48*, 153–161. [CrossRef]
80. Gomes, S.V.; Portugal, L.A.; dos Anjos, J.P.; de Jesus, O.N.; de Oliveira, E.J.; David, J.P.; David, J.M. Accelerated solvent extraction of phenolic compounds exploiting a Box–Behnken design and quantification of five flavonoids by HPLC-DAD in *Passiflora* species. *Microchem. J.* **2017**, *132*, 28–35. [CrossRef]
81. Guo, L.; Zhu, W.; Xu, F.; Liu, M.; Xie, Y.; Zhang, J. Optimized ultrasonic-assisted extraction of polysaccharides from *Cyclina sinensis* and evaluation of antioxidant activities in vitro. *CyTA J. Food* **2014**, *12*, 32–39. [CrossRef]
82. Waris, M.; Kocak, E.; Gonulalan, E.M.; Demirezer, L.O.; Kır, S.; Nemutlu, E. Metabolomics analysis insight into medicinal plant science. *TrAC Trends Anal. Chem.* **2022**, *157*, 116795. [CrossRef]
83. Lee, J.; Ji, J.; Park, S.H. Antiwrinkle and antimelanogenesis activity of the ethanol extracts of *Lespedeza cuneata* G. Don for development of the cosmeceutical ingredients. *Food Sci. Nutr.* **2018**, *6*, 1307–1316. [CrossRef]
84. Lai, Z.; Tsugawa, H.; Wohlgemuth, G.; Mehta, S.; Mueller, M.; Zheng, Y.; Ogiwara, A.; Meissen, J.; Showalter, M.; Takeuchi, K.; et al. Identifying metabolites by integrating metabolome databases with mass spectrometry cheminformatics. *Nat. Methods* **2018**, *15*, 53–56. [CrossRef] [PubMed]

molecules

MDPI

Article

Bioactive Power of Black Chokeberry Pomace as Affected by Advanced Extraction Techniques and Cryogrinding

Maja Repajić, Marija Zorić, Ivan Magnabosca, Sandra Pedisić, Verica Dragović-Uzelac and Ivona Elez Garofulić *

University of Zagreb Faculty of Food Technology and Biotechnology, Pierottijeva 6, 10000 Zagreb, Croatia; maja.repajic@pbf.unizg.hr (M.R.); sandra.pedisic@pbf.unizg.hr (S.P.); verica.dragovic-uzelac@pbf.unizg.hr (V.D.-U.)

* Correspondence: ivona.elez@pbf.unizg.hr

Abstract: Black chokeberry (*Aronia melanocarpa* L.) pomace (BCP), a major by-product of juice production, is an underutilized source of polyphenols and anthocyanins with strong antioxidant properties. This study aimed to optimize and compare three green extraction techniques—pressurized liquid extraction (PLE), microwave-assisted extraction (MAE), and ultrasound-assisted extraction (UAE)—for recovering total polyphenols (TP) and total monomeric anthocyanins (TMA) from BCP, with reflux extraction as a benchmark. The effects of temperature, extraction time, and solid–solvent ratio were evaluated, and cryogrinding was assessed as a pre-treatment. PLE achieved the highest TP yields at elevated temperatures but reduced anthocyanin recovery, while MAE offered a balance of high TP and TMA, with strong antioxidant capacity. Cryogrinding enhanced TP extraction, with only 1 min of cryogrinding maximizing yield. UPLC-MS/MS analysis of optimized MAE extract confirmed cyanidin-3-glucoside and cyanidin-3-galactoside as dominant anthocyanins, alongside notable flavonols and phenolic acids, validating the rich phenolic profile. Overall, MAE combined with 1 min of cryogrinding proved to be the most effective approach for preserving heat-sensitive compounds while achieving high yields. These findings demonstrate that optimized green extraction can efficiently valorize BCP, supporting sustainable food processing and waste reduction in line with circular economy principles.

Keywords: *Aronia melanocarpa* L.; valorization; PLE; MAE; UAE; optimization; polyphenols; anthocyanins; antioxidant; cryomilling

1. Introduction

The increasing consumer demand for natural antioxidants and functional ingredients has intensified interest in the sustainable sourcing of bioactive compounds from food industry by-products. Among these, black chokeberry (*Aronia melanocarpa* L.) stands out as a unique raw material due to its exceptionally high content of polyphenolic compounds, particularly anthocyanins, proanthocyanidins, phenolic acids, and flavonols. These compounds are well documented for their strong antioxidant activity, radical scavenging properties, and associated health benefits, including anti-inflammatory, cardioprotective, and anti-carcinogenic effects [1].

Black chokeberry is widely processed into juices, syrups, and concentrates, generating substantial amounts of pomace as a by-product. This pomace consists predominantly of berry skins, seeds, and residual pulp and retains a considerable proportion of the fruit's original phenolic content. According to Kaloudi et al. [2], only 13% of the fruit's total phenolic content is transferred to the juice during pressing, 10% of the total anthocyanins, 66% of the phenolic acids, and 23% of the flavonoids, indicating that a considerable amount, comprising predominantly anthocyanins and flavonoids, remains in the pomace and could be further extracted for valorization. Indeed, previous studies have characterized black chokeberry pomace (BCP) as the main source of anthocyanins such as cyanidin-3-galactoside and cyanidin-3-glucoside, flavonols (e.g., quercetin and kaempferol glycosides), and phenolic acids (e.g., chlorogenic and neochlorogenic acids), which contribute synergistically to its antioxidant potential [2–5].

Despite its high value, BCP remains largely underutilized and is often discarded or used for low-value applications such as animal feed or compost. In the context of sustainable production and circular economy principles, the valorization of this bio-residue into valuable antioxidant extracts represents a promising strategy to reduce food waste and add economic value to processing chains. However, the effective recovery of these sensitive compounds depends greatly on the choice of extraction method and the optimization of critical process parameters.

Traditional extraction approaches, including reflux or maceration, typically require prolonged extraction times, elevated temperatures, and high solvent consumption, which can lead to the degradation of heat-sensitive anthocyanins and increase energy costs [6]. In contrast, green extraction technologies such as pressurized liquid extraction (PLE), microwave-assisted extraction (MAE), and ultrasound-assisted extraction (UAE) offer significant advantages by enhancing extraction efficiency, reducing process time and solvent usage, and minimizing thermal degradation of target compounds [7]. The efficiency of these advanced methods, however, is highly dependent on the precise control of key operational parameters. For instance, temperature has been shown to play a dual role: elevated temperatures can improve the solubility and mass transfer of phenolics, but may simultaneously degrade thermolabile anthocyanins, as demonstrated in studies on grape and blackberry pomace [8,9]. Similarly, extraction time must be carefully optimized to balance sufficient cell wall disruption with minimal compound degradation. The solid–solvent ratio (SSR) is another critical factor; increasing the SSR generally enhances extraction efficiency by providing more solvent per unit of sample, thus improving diffusion rates and compound solubilization until extraction equilibrium is reached [10].

Furthermore, mechanical pre-treatments such as cryogrinding have gained attention as simple yet effective methods to increase extraction yields by reducing particle size, which increases the available surface area for mass transfer and facilitates solvent penetration into plant matrices [11]. Recent studies have shown that cryogrinding can significantly reduce particle size and enhance phenolic recovery from myrtle leaves [12], fenugreek seeds [13], and sword beans [14].

Although the separate application of green extraction techniques and mechanical pre-treatments has been widely explored, comprehensive studies comparing the combined effects of optimized PLE, MAE, and UAE parameters, together with cryogrinding, for maximizing both total polyphenols and anthocyanins from BCP remain scarce.

Therefore, the aim of this study was to systematically investigate the influence of critical extraction parameters, temperature, extraction time, and SSR on the recovery of total polyphenols and total monomeric anthocyanins from BCP using PLE, MAE, and UAE. Furthermore, the study assessed the additional impact of cryogrinding as a pre-treatment to maximize extraction efficiency. The resulting extract was characterized for individual phenolic composition, and the antioxidant capacity was evaluated through multiple assays to provide a comprehensive understanding of its functional potential. By optimizing and comparing these techniques, this work seeks to provide a practical basis for the sustainable valorization of BCP, demonstrating its potential as a high-value source of natural antioxidants and contributing to the broader goals of food waste reduction and a circular bioeconomy.

2. Results and Discussion

2.1. Determination of PLE, MAE, and UAE Optimal Process Conditions

The results of total polyphenol (TP) and total monomeric anthocyanin (TMA) content determination in BCP extracts obtained under different process conditions of PLE, MAE, and UAE are given in Table 1.

The TP content ranged from 86.0 to 132.5 mg gallic acid equivalents (GAE)/g of dry matter (dm) under PLE conditions, from 58.5 to 95.7 mg GAE/g dm under MAE conditions, and from 63.1 to 88.9 mg GAE/g dm under UAE conditions. Woźniak et al. [15] reported a TP content ranging from 3.67 to 15.21 mg GAE/g of thawed chokeberry pomace, using supercritical carbon dioxide extraction with ethanol under varying conditions of temperature, pressure, and ethanol concentration. In comparison, the TP values obtained in this study using PLE were 88–96% higher, indicating that PLE is a more efficient method for extracting phenolic compounds from BCP than supercritical carbon dioxide extraction. Simić et al. [16] reported TP contents ranging from 3.73 to 4.58 mg GAE/g of frozen chokeberries when studying the impact of MAE exposure time from 5 to 15 min and microwave power ranging from 300–600 W on TP content in chokeberry fruit, which is significantly lower (94–95%) than the TP content found in this study. However, the mentioned study employed an extraction approach with no temperature control, so excessive degradation of heat sensitive compounds could have occurred. It should also be noted that none of the above studies used dried chokeberry material. As for UAE, slightly comparable values (16.85 to 74.28 mg GAE/g dm, i.e., 16–73% lower) to those obtained in this work were reported by D'Alessandro et al. [17], who investigated the TP content in dried BCP extracts obtained by UAE while evaluating the influence of extraction time (0–240 min), temperature (20–70 °C), solvent composition (0–50% ethanol in water), and ultrasound power (0–100 W).

Observed ranges for TMA content were 8.2–24.1 mg cyanidin-3-glucoside equivalents (C3GE)/g dm in PLE extracts, 20.3–26.6 mg C3GE/g dm in MAE extracts, and 18.7–23.2 mg C3GE/g dm in UAE extracts. Roda-Serrat et al. [18] reported anthocyanin contents of 18.1–51.0 mg/dry weight in chokeberry pomace using homogenization in citric acid aqueous solution (0.25, 0.75, and 1.5%, *w*/*v*), which is 53–55% higher compared to the PLE yields obtained in this study. Elez Garofulić et al. [3] reported 79–83% lower anthocyanin contents (3.35–5.53 mg/g) in thawed BCP pomace using MAE under various irradiation time (4, 6, 8, and 10 min) and temperature (40, 60, and 80 °C) combinations. They also applied UAE under different amplitudes (25, 50, and 75%) and sonication times (4, 6, 8, and 10 min), achieving TMA yields of 3.89–5.67 mg/g, which is 75–79% lower than the UAE yields obtained in this study. In addition to the use of undried material, their high SSR (1:100 g/mL) may have limited extraction efficiency compared to the values obtained under MAE and UAE in this study.

The influence of individual PLE, MAE, and UAE process parameters on TP and TMA in BCP extracts is shown in Table 2. The presented results indicate that temperature had a statistically significant effect on TP and TMA content during PLE, while there was no observed effect of static extraction time and SSR. The highest average value of TP content was achieved in the extract obtained at 150 °C, while there was no statistically significant difference between the TP content extracted at 100 and 125 °C. This suggests that the successful extraction of phenolic compounds from BCP favors the use of higher temperatures. Pereira et al. [8] conducted PLE on grape pomace at temperatures of 40, 60, 80, and 100 °C using different solvents and an extraction time of 4 h. Using a 50% aqueous ethanol solution as the extraction solvent, they observed that increasing the temperature resulted in a significant increase in TP content. Similarly, Dobroslavić et al. [19] observed an increase in TP content with increasing temperature. They performed PLE on bay leaves at 90, 120, and 150 °C and concluded that the TP content increased proportionally with the rise in temperature. However, in the extraction of anthocyanins, the opposite trend can be observed. The highest average content of TMA was recorded in the extract obtained at 100 °C, and with each further increase in temperature, their content decreased, with the lowest value observed in the extract obtained at 150 °C. According to the previously mentioned study by Pereira et al. [8], increasing the PLE temperature from 40 to 100 °C also resulted in a decrease in the TMA content in grape pomace extracts.

This trend suggests that raising the temperature above 100 °C negatively affects the yield of anthocyanins, which is understandable considering that anthocyanins are thermolabile and degrade when exposed to elevated temperatures. Similar results were reported by Ju and Howard [20] in their study, in which they conducted PLE on dried grape skin at temperatures ranging from 20 to 140 °C using various solvents. When a 60% aqueous ethanol solution containing 0.1% hydrochloric acid was used as an extraction solvent, the TMA content decreased with increasing temperature, while the TP content increased. Da Fonseca Machado et al. [9] performed PLE on blackberry pomace using different solvents at 60, 80, and 100 °C. In their study, they also observed that higher temperatures led to an increase in TP content, while the TMA content decreased. The difference between the loss of anthocyanins at elevated extraction temperatures and the increased yield of TP was attributed to the enhanced extraction of more thermostable compounds, such as procyanidins and phenolic acids.

Considering the obtained results and observed effects, it was concluded that optimal PLE process conditions for the simultaneous extraction of both TP and TMA were temperature 125 °C, static time 5 min, and SSR 1:40 g/mL.

During the MAE, temperature and extraction time significantly affected the TP content of BCP extracts, with no observed effect on anthocyanins, while the SSR significantly affected both TP and TMA content. The highest average value of TP content was achieved at 80 °C. This value was statistically significantly higher than that obtained at 40 °C, but did not differ significantly from the value obtained at 60 °C, indicating that further temperature increases above 60 °C did not have a significant effect on TP content. Wu et al. [21] investigated the use of MAE for isolating phenolic compounds from potatoes at temperatures of 50, 60, 70, and 80 °C. They observed a significant increase in TP concentration with rising temperature. A similar pattern was reported by Jokić et al. [22] during the optimization of MAE for phenolic extraction from broccoli. In their study, the concentration of extracted phenolics increased as the temperature rose from 50 to 73 °C, while further heating to 90 °C resulted in a decrease in TP content. Zhao et al. [23] also observed a comparable trend when applying MAE to *Melastoma sanguineum*. Increasing the temperature from 20 to 50 °C led to a significant rise in TP content, whereas a further increase to 70 °C did not produce a statistically significant change.

Furthermore, extending the irradiation time from 5 to 10 min resulted in an increase in TP content. As previously mentioned, Zhao et al. [23] also examined the effect of MAE irradiation time on TP yield. Although their extraction times were considerably longer than those used in this study, they similarly observed that extending the irradiation time from 15 to 45 min positively influenced TP yield. However, further prolongation to 60 min led to a slight decrease in TP content.

Table 1. Content of TP and TMA in BCP extracts obtained under PLE, MAE, and UAE.

Extraction Technique	Temperature (°C)	Amplitude (%)	Time (min)	SSR (g/mL)	TP (mg GAE/g dm)	TMA (mg C3GE/g dm)
PLE	100		5	1:40	86.0 ± 0.2	22.5 ± 0.1
	100		5	1:60	101.6 ± 1.6	24.1 ± 0.1
	100		5	1:80	113.1 ± 4.6	23.9 ± 0.3
	100		10	1:40	99.2 ± 4.6	22.4 ± 0.1
	100		10	1:60	108.6 ± 5.6	21.5 ± 0.5
	100		10	1:80	118.1 ± 6.7	20.6 ± 0.1
	125		5	1:40	101.3 ± 8.7	19.3 ± 0.2
	125		5	1:60	97.6 ± 3.0	18.2 ± 0.3
	125		5	1:80	115.6 ± 0.8	18.7 ± 0.1
	125		10	1:40	110.4 ± 4.4	15.0 ± 0.1
	125		10	1:60	119.2 ± 3.4	15.6 ± 0.4
	125		10	1:80	108.7 ± 1.0	16.2 ± 0.1
	150		5	1:40	119.8 ± 1.3	13.7 ± 0.3
	150		5	1:60	123.1 ± 5.0	14.1 ± 0.1
	150		5	1:80	128.7 ± 0.2	13.4 ± 0.1
	150		10	1:40	126.2 ± 4.2	12.0 ± 0.1
	150		10	1:60	123.9 ± 3.9	12.3 ± 0.2
	150		10	1:80	132.5 ± 7.4	8.2 ± 0.1
MAE	40		5	1:40	58.5 ± 1.5	20.7 ± 0.4
	40		5	1:60	69.3 ± 5.9	23.7 ± 0.2
	40		5	1:80	75.5 ± 1.0	25.9 ± 0.7
	40		10	1:40	63.5 ± 0.4	23.1 ± 1.3
	40		10	1:60	72.7 ± 2.1	24.3 ± 0.8
	40		10	1:80	79.8 ± 1.8	26.6 ± 0.1
	60		5	1:40	65.4 ± 1.7	20.8± 0.1
	60		5	1:60	72.8 ± 1.8	24.5 ± 0.5
	60		5	1:80	81.3 ± 4.5	24.5 ± 0.1
	60		10	1:40	70.6 ± 0.1	22.7 ± 0.2
	60		10	1:60	75.5 ± 2.7	23.8 ± 0.5
	60		10	1:80	90.1 ± 4.3	25.6 ± 0.2
	80		5	1:40	68.9 ± 0.8	20.3 ± 0.2
	80		5	1:60	84.7 ± 2.0	22.3 ± 0.5
	80		5	1:80	82.4 ± 6.7	24.8 ± 0.1
	80		10	1:40	76.1 ± 1.2	22.7 ± 0.2
	80		10	1:60	95.7± 6.52	25.2 ± 0.4
	80		10	1:80	94.8 ± 8.3	24.0 ± 0.5
UAE		50	5	1:40	65.2 ± 2.2	20.9 ± 0.7
		50	5	1:60	79.9 ± 2.7	23.2 ± 0.9
		50	5	1:80	75.2 ± 1.5	22.4 ± 0.4
		50	10	1:40	69.7 ± 1.0	20.8 ± 0.1
		50	10	1:60	74.4 ± 4.3	20.4 ± 0.1
		50	10	1:80	73.9 ± 1.8	20.5 ± 0.5
		75	5	1:40	70.6 ±2.8	21.7 ± 0.2
		75	5	1:60	75.3 ± 6.6	21.0 ± 0.1
		75	5	1:80	74.3 ± 1.2	21.5 ± 0.3
		75	10	1:40	66.7 ± 1.0	21.0 ± 0.1
		75	10	1:60	71.7 ± 1.6	22.6 ± 0.3
		75	10	1:80	84.1 ± 2.2	21. 9 ± 0.1
		100	5	1:40	64.8 ± 2.2	20.2 ± 0.1
		100	5	1:60	75.7 ± 3.6	22.5 ± 0.4
		100	5	1:80	74.6 ± 5.7	21.3 ± 0.1
		100	10	1:40	63.1 ± 0.4	19.9 ± 1.3
		100	10	1:60	69.8 ± 5.7	18.7 ± 0.2
		100	10	1:80	88.9 ± 2.0	22.2 ± 0.1

TP—total polyphenols; TMA—total monomeric anthocyanins; BCP—black chokeberry pomace; PLE—pressurized liquid extraction; MAE—microwave-assisted extraction; SSR—solid–solvent ratio; GAE—gallic acid equivalents; C3GE—cyanidin-3-glucoside equivalents. Results are expressed as mean ± standard deviation (SD).

The SSR had a statistically significant effect on all observed dependent variables. The highest average values for TP and TMA content were achieved at an SSR of 1:80 g/mL. The lowest values for both variables were recorded at a ratio of 1:40 g/mL. However, statistical analysis revealed no significant difference in TP content between the 1:60 and 1:80 g/mL ratios, suggesting that a 1:60 g/mL ratio is sufficient for effective extraction of TP from BCP using MAE. In contrast, for maximum TMA yield, the highest SSR (1:80 g/mL) should be considered. A similar trend was observed in the study by Wen et al. [24], who applied

MAE to blackberry fruits. They reported that increasing the SSR (1:15, 1:20, and 1:25 g/mL) led to an increase in TP content. Pap et al. [25] also found that increasing the SSR (from 1:10 to 1:20 g/mL) positively influenced anthocyanin extraction from black currants using MAE. Likewise, Zhao et al. [23], in their MAE study on *M. sanguineum*, recorded a significant increase in TP content with increasing SSR (1:10, 1:20, and 1:30 g/mL). However, further increases to 1:40, 1:50, and 1:60 g/mL led to a slight decrease in phenolic yield, suggesting that extraction equilibrium in mass transfer was likely reached at a ratio of 1:30 g/mL.

Considering the statistical analysis of all examined dependent variables, the optimal conditions for efficient MAE of TP and TMA from BCP were determined to be a temperature of 80 °C, an extraction time of 10 min, and an SSR of 1:80 g/mL.

When applying UAE as an extraction method, amplitude did not exhibit a statistically significant effect on any of the observed dependent variables, with values remaining consistent across all amplitudes tested. In line with the findings of Elez Garofulić et al. [3], who also investigated UAE from BCP, increasing the amplitude from 25 to 75% did not result in a statistically significant change in the yield of TMA. Similarly, Borrás-Enríquez et al. [26], who applied UAE to mango by-products, reported no significant differences in TP content across amplitudes ranging from 30 to 90%.

Extraction time demonstrated a statistically significant effect only on the content of TMA, with a shorter extraction duration (5 min), proving more effective for anthocyanin recovery. Comparable findings were reported by González et al. [27], who examined the impact of extraction time (2–25 min) on anthocyanin yield from blackcurrant using UAE. Their results also indicated that a shorter extraction time (5 min) yielded higher anthocyanin content, whereas extended durations (10–20 min) led to reduced anthocyanin yields, likely due to degradation.

The SSR had a statistically significant influence on TP content, with no effect on TMA. The highest average TP content was achieved at SSRs of 1:60 and 1:80 g/mL, while the lowest was recorded at 1:40 g/mL. These results indicate that increasing solvent volume enhances phenolic extraction efficiency. However, further increases beyond a 1:60 g/mL ratio did not result in statistically significant improvements in phenolic yield. A similar trend was observed by He et al. [28] during UAE of blueberry pomace, where increasing the SSR from 1:15 to 1:20 g/mL improved TP content, but a ratio of 1:25 g/mL did not yield a significant additional benefit. D'Alessandro et al. [17] reported comparable results for UAE applied to BCP, where increasing the SSR from 1:10 to 1:20 g/mL enhanced phenolic recovery, but a further increase to 1:40 g/mL did not lead to a significant improvement in yield.

Considering all the observations above, optimal UAE parameters for simultaneous recovery of TP and TMA were set at amplitude 50%, extraction time 5 min, and SSR 1:60 g/mL.

Finally, to examine the possible grouping of the obtained BCP extracts according to the content of TP and TMA with respect to the applied extraction technique, principal component analysis (PCA) was performed. The results are shown in Figure 1.

The communality values of TP and TMA exceeded 0.8, indicating that both variables are well represented by the extracted components and contribute significantly to the factor structure. The distribution of the samples was examined in the space defined by the first two principal components (PC1 and PC2), both of which had eigenvalues greater than 1. Together, PC1 and PC2 explained 100.00% of the total variance in the data. A very strong correlation was observed between PC1 and TP ($r > 0.9$), as well as between PC1 and TMA ($r < -0.9$). In contrast, moderate correlations ($r > 0.4$) were found between PC2 and both TP and TMA. The results reveal a clear grouping of extract samples according to the extraction technique. Almost all samples obtained using PLE were located in the area of positive PC1 values and were characterized by high TP content. In contrast, all samples obtained using MAE and UAE clustered in the area of negative PC1 values, characterized by high levels of TMA. These results confirm earlier observations showing clear differences in the efficiency of PLE, MAE, and UAE in isolating phenolic compounds and anthocyanins from BCP, with PLE demonstrating the highest efficiency in the isolation of phenolic compounds. However, the high temperature applied during PLE (≥ 100 °C) is unfavorable for anthocyanin isolation due to its thermolability, which justifies the use of MAE. The milder conditions applied during MAE, particularly the temperature (max. 80 °C), preserved anthocyanins while still achieving a satisfactory yield of phenolic compounds. Therefore, MAE can be highlighted as an effective method for obtaining BCP extracts rich in both phenolic compounds and anthocyanins.

Table 2. The influence of PLE, MAE, and UAE process conditions on TP and TMA in BCP extracts.

Source of Variation	TP (mg GAE/g dm)	TMA (mg C3GE/g dm)
	PLE	
Temperature (°C)	$p < 0.001$ *	$p < 0.001$ *
100	104.4 ± 3.3 [a]	22.5 ± 0.4 [c]
125	108.8 ± 2.5 [a]	17.3 ± 0.5 [b]
150	125.7 ± 1.5 [b]	12.3 ± 0.6 [a]
Time (min)	$p = 0.155$	$p = 0.070$
5	109.6 ± 3.2 [a]	18.7 ± 1.0 [a]
10	116.3 ± 2.5 [a]	16.0 ± 1.1 [a]
SSR (g/mL)	$p = 0.053$	$p = 0.953$
1:40	107.2 ± 4.2 [a]	17.5 ± 1.3 [a]
1:60	112.3 ± 3.2 [a]	17.7 ± 1.3 [a]
1:80	119.4 ± 2.7 [a]	16.8 ± 1.5 [a]
	MAE	
Temperature (°C)	$p = 0.003$ *	$p = 0.552$
40	69.9 ± 2.2 [a]	24.0 ± 0.6 [a]
60	76.0 ± 2.5 [ab]	23.6 ± 0.5 [a]
80	83.8 ± 3.1 [b]	23.2 ± 0.5 [a]
Time (min)	$p = 0.050$ *	$p = 0.054$
5	73.2 ± 2.1 [a]	23.1 ± 0.5 [a]
10	79.9 ± 2.7 [b]	24.2 ± 0.3 [a]
SSR (g/mL)	$p < 0.001$ *	$p < 0.001$ *
1:40	67.2 ± 1.7 [a]	21.7 ± 0.4 [a]
1:60	78.5 ± 2.9 [b]	24.0 ± 0.3 [b]
1:80	84.0 ± 2.2 [b]	25.2 ± 0.3 [c]
	UAE	
Amplitude (%)	$p = 0.944$	$p = 0.228$
50	73.1 ± 1.5 [a]	21.4 ± 0.3 [a]
75	73.8 ± 1.8 [a]	21.6 ± 0.2 [a]
100	72.8 ± 2.7 [a]	20.8 ± 0.4 [a]
Time (min)	$p = 0.749$	$p = 0.050$ *
5	72.9 ± 1.3 [a]	21.6 ± 0.2 [b]
10	73.6 ± 1.9 [a]	20.9 ± 0.3 [a]
SSR (g/mL)	$p < 0.001$ *	$p = 0.161$
1:40	66.7 ± 0.9 [a]	20.7 ± 0.2 [a]
1:60	74.5 ± 1.4 [b]	21.4 ± 0.5 [a]
1:80	78.5 ± 1.9 [b]	21.6 ± 0.2 [a]

PLE—pressurized liquid extraction; MAE—microwave-assisted extraction; UAE—ultrasound-assisted extraction; TP—total polyphenols; TMA—total monomeric anthocyanins; BCP—black chokeberry pomace; SSR—solid-solvent ratio; GAE—gallic acid equivalents; C3GE—cyanidin-3-glucoside equivalents. Results are expressed as mean ± standard error (SE). * $p \leq 0.05$. Means with the same letter within a column are not significantly different at $p \leq 0.05$.

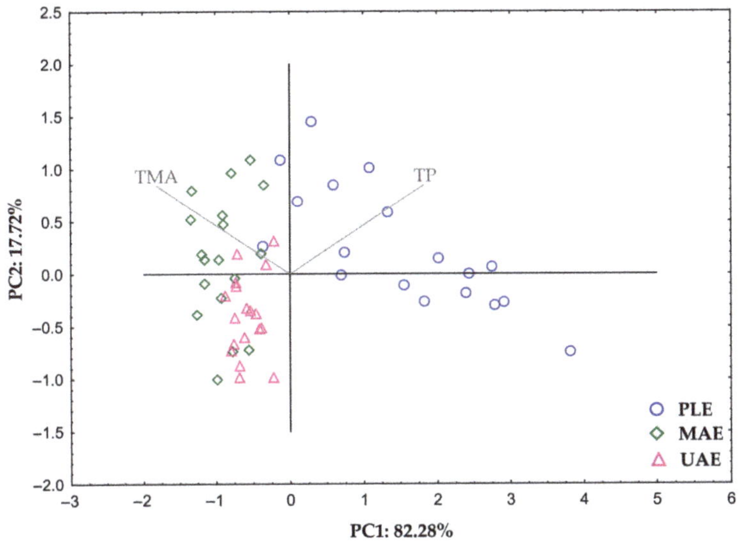

Figure 1. Distribution of BCP extracts in a two-dimensional coordinate system defined by the first two principal components (PC1 and PC2) in relation to the type of extraction technique (BCP—black chokeberry pomace; PLE—pressurized liquid extraction; MAE—microwave-assisted extraction; UAE—ultrasound-assisted extraction; TP—total polyphenols; TMA—total monomeric anthocyanins).

2.2. *Efficiency Comparison of PLE, MAE, UAE, and Reflux*

Optimized extraction techniques (PLE, MAE, and UAE) were compared with reflux as a conventional extraction method in terms of TP and TMA yield and antioxidant capacity of obtained extracts determined by multiple assays (FRAP, DPPH, ABTS, and ORAC), each capturing different mechanisms of antioxidant activity (Table 3).

Table 3. Content of TP, TMA, and antioxidant capacity in BCP extracts obtained under PLE, MAE, and UAE optimal conditions and reflux extraction.

Extraction Technique	TP (mg GAE/g dm)	TMA (mg C3GE/g dm)	FRAP (μmol TE/g dm)	DPPH (μmol TE/g dm)	ABTS (μmol TE/g dm)	ORAC (μmol TE/g dm)
	$p = 0.012$ *	$p = 0.007$ *	$p = 0.002$ *	$p = 0.001$ *	$p < 0.001$ *	$p < 0.001$ *
PLE	101.3 ± 8.7 [ab]	19.3 ± 0.2 [a]	1052.1 ± 48.9 [b]	297.1 ± 2.5 [a]	519.2 ± 9.4 [a]	92.2 ± 2.2 [a]
MAE	94.8 ± 8.3 [a]	24.0 ± 0.5 [b]	983.7 ± 60.5 [b]	420.4 ± 8.0 [b]	681.3 ± 24.8 [b]	199.6 ± 1.6 [c]
UAE	79.9 ± 2.7 [a]	23.2 ± 0.9 [b]	637.4 ± 28.1 [a]	315.7 ± 3.2 [a]	461.3 ± 15.0 [a]	142.7 ± 1.9 [b]
REFLUX	120.8 ± 0.8 [b]	22.9 ± 0.8 [b]	1033.4 ± 16.7 [b]	472.9 ± 25.3 [b]	790.2 ± 14.4 [c]	222.6 ± 2.9 [d]

TP—total polyphenols; TMA—total monomeric anthocyanins; BCP—black chokeberry pomace; PLE—pressurized liquid extraction; MAE—microwave-assisted extraction; UAE—ultrasound-assisted extraction; GAE—gallic acid equivalents; C3GE—cyanidin-3-glucoside equivalents; TE—Trolox equivalents. Results are expressed as mean ± SD. * $p \leq 0.05$. Means with the same letter within a column are not significantly different at $p \leq 0.05$.

Reflux extraction yielded the highest TP content (120.8 mg GAE/g dm), followed by PLE and MAE, with 101.3 and 94.8 mg GAE/g dm, respectively, indicating a limitation of ultrasound-based methods for phenolic recovery under the tested conditions. These results are consistent with previous reports where thermal methods enhanced phenolic solubility and matrix softening, leading to increased extraction yields [29].

However, the highest TMA content was recorded in MAE extracts (24.0 mg C3GE/g dm), followed closely by UAE and reflux, while PLE yielded significantly less (19.3 mg C3GE/g dm). This suggests that MAE may be particularly effective for extracting thermolabile anthocyanins due to shorter processing times and lower temperatures, unlike more intensive processes such as PLE. This also aligns with previous studies indicating that microwave heating favors the extraction of anthocyanins due to rapid cell wall rupture and reduced exposure to elevated temperatures [6].

Regarding antioxidant capacity, different trends were observed in relation to the applied assay. The FRAP assay, which reflects electron-donating ability, showed that PLE (1052.1 μmol TE/g dm) and reflux (1033.4 μmol TE/g dm) extractions were superior, with no significant difference between them. These results correlate with their high TP content, underscoring the strong reducing capacity of polyphenol-rich extracts. MAE (983.7 μmol TE/g dm) also showed comparable FRAP values, while UAE was significantly lower (637.4 μmol TE/g dm), aligning with its reduced TP content. In the DPPH assay, which is also based on single electron transfer, MAE and reflux outperformed PLE and UAE, yielding values of 420.4 and 472.9 μmol TE/g dm, respectively. The ABTS assay revealed significantly higher values in MAE (681.3 μmol TE/g dm) and reflux (790.2 μmol TE/g dm) extracts. PLE and UAE showed moderate ABTS values (519.2 and 461.3 μmol TE/g dm, respectively). This trend in ABTS values may only be partially related to TMA content. The literature data supports the observation that anthocyanins, particularly cyanidin derivatives, are efficient ABTS radical scavengers due to their structural features, specifically hydroxylation and conjugation patterns that stabilize radical intermediates [30]. However, the results obtained indicate a possible interference of other non-phenolic compounds in the ABTS assay. Some of the possible interactions could be those of lipophilic compounds present in BCP that to a certain extent partition into alcohol-based solvents during the extraction with ABTS radical, as ABTS assay is known to detect both hydrophilic and some lipophilic antioxidants, unlike the FRAP, DPPH, and ORAC assay, when performed in the hydrophilic mode [31]. The ORAC results followed a trend similar to ABTS: reflux extraction yielded the highest ORAC value (222.6 μmol TE/g dm), followed closely by MAE (199.6 μmol TE/g dm), both significantly surpassing UAE (142.7 μmol TE/g dm) and especially PLE (92.2 μmol TE/g dm). These findings indicate that conventional and microwave-based methods may preserve peroxyl radical-scavenging capacity better than pressurized and ultrasound-assisted methods, possibly due to less structural degradation of antioxidant molecules. This observation can also be attributed to TMA content. Anthocyanins, particularly cyanidin derivatives, are well-known to contribute significantly to ORAC values because their ortho-dihydroxy B-ring structure and conjugated system make them good hydrogen donors. This structural feature makes them more reactive in HAT-based assays (like ORAC) than in some SET-based assays, such as FRAPP and DPPH [32]. Accordingly, Wu et al. [33] showed that black chokeberry's high ORAC values are strongly associated with its high levels of cyanidin glycosides.

In summary, while PLE maximized TP yield and demonstrated strong electron transfer capacity in FRAP assay similar to conventional reflux extraction, MAE proved to be the most balanced advanced technique overall in relation to TP and TMA content and antioxidant capacity.

2.3. *Effects of Cryogrinding*

In order to evaluate cryogrinding as a pre-treatment for maximizing the extraction yield, samples of BCP were ground using a conventional method, marked as 0 min cryogrinding time, as well as by cryogrinding for different durations, namely 1, 3, 5, 7, and 9 min. The results of the particle size distribution measurement are shown in Table 4.

Table 4. Particle size of BCP powder obtained by conventional grinding and cryogrinding.

Cryogrinding Time (min)	d(0.1) (μm)	d(0.5) (μm)	d(0.9) (μm)	Span
	$p < 0.001$ *	$p < 0.001$ *	$p < 0.001$ *	$p < 0.001$ *
0	37.0 ± 2.8 ^b^	238.1 ± 10.9 ^d^	712.9 ± 43.4 ^c^	2.8 ± 0.0 ^a^
1	5.0 ± 0.2 ^a^	75.3 ± 2.3 ^c^	455.6 ± 17.3 ^b^	6.0 ± 0.1 ^b^
3	3.7 ± 0.1 ^a^	53.1 ± 4.2 ^b^	419.7 ± 70.1 ^b^	7.8 ± 0.7 ^bc^
5	2.6 ± 0.1 ^a^	24.9 ± 1.1 ^a^	245.1 ± 29.2 ^a^	9.7 ± 0.7 ^c^
7	2.3 ± 0.1 ^a^	19.4 ± 0.7 ^a^	186.7 ± 19.8 ^a^	9.5 ± 0.7 ^c^
9	2.1 ± 0.1 ^a^	15.4 ± 0.8 ^a^	135.6 ± 13.5 ^a^	8.7 ± 0.5 ^c^

BCP—black chokeberry pomace. Results are expressed as mean ± SD. * $p \leq 0.05$. Means with the same letter within a column are not significantly different at $p \leq 0.05$.

The highest value for the d(0.1) parameter was recorded in the sample ground using the conventional grinding method, while the lowest value was observed in the sample after 9 min of cryogrinding. It is evident that the d(0.1) values decrease with increasing grinding time, demonstrating the effectiveness of cryogrinding in reducing the size of 10% of the particles. Notably, just 1 min of cryogrinding was sufficient to reduce 10% of the particles by 86.5% compared to conventional grinding. However, further extension of grinding time did not result in statistically significant differences in d(0.1) values among the cryogenically ground samples. Similar findings were reported by Kraljić et al. [11], who performed cryogrinding of rapeseed press cake for 2, 4, 8, and 12 min. The lowest d(0.1) value was likewise recorded in the sample ground for the longest duration (12 min), resulting in a reduction of 10% of the particles by 89.23% compared to the conventionally ground sample. Regarding the values for d(0.5), the highest value was again found in the conventionally ground sample, while the lowest was observed in the sample after 9 min of cryogrinding. A statistically significant difference in the d(0.5) parameter was found between the conventionally ground sample and those cryogenically ground for 1, 3, and 5 min, with the 50% particle size being reduced by 68.4%, 77.7%, and even 89.5% after 1, 3, and 5 min of cryogrinding, respectively. Further extension of the grinding time to 7 or 9 min did not lead to a significant additional reduction in the 50% particle size. Kraljić et al. [11] also reported a decrease in d(0.5) values with prolonged cryogrinding, observing statistically significant differences among all samples compared to the conventional sample, with reductions in the 50% particle size of 56.28, 65.09, 76.55, and 84.30% after 2, 4, 8, and 12 min of cryogrinding, respectively. The 90% particle size (d(0.9)) followed the same trend as the previously discussed parameters: the highest value was recorded in the conventionally ground sample, and the lowest in the sample ground cryogenically for 9 min. Statistical analysis showed results similar to those for d(0.5), except that no significant difference was observed in the 90% particle size for the samples ground cryogenically for 1 and 3 min. The results indicate that the reduction in the 90% particle size was 36.0% and 41.1% after 1 and 3 min of cryogrinding compared to conventional grinding, while 5 min of cryogrinding reduced the 90% particle size by 65.6%. Further extension of cryogrinding to 7 and 9 min did not contribute to a statistically significant additional reduction in the 90% particle size. In contrast to the d(0.9) values obtained in this study, all d(0.9) values in the work by Kraljić et al. [11] showed statistically significant differences. They recorded the lowest value in the sample subjected to the longest cryogrinding (12 min), with a reduction of 90% of the particles by 83.12% compared to the conventionally ground sample.

The final parameter examined was the particle size distribution span, which describes the homogeneity of the particle size distribution and should be at the lowest values, thereby indicating good uniformity of the particle size. The calculated span values ranged from 2.8 to 9.7 μm, with the highest span recorded in the sample after 5 min of cryogenic grinding and the lowest in the conventionally ground sample. Given the higher span values in the cryogenically ground samples, it can be concluded that the particle distribution in these samples was less homogeneous, whereas the greatest uniformity was observed in the conventionally ground sample. Kraljić et al. [11] reported the lowest span value in the sample obtained after 4 min of cryogenic grinding, indicating the highest uniformity in that sample. Conversely, the lowest homogeneity was observed in the sample ground for the longest duration (12 min), which was also confirmed in this study.

Considering all the particle size parameters discussed, it can be concluded that cryogrinding is highly effective in significantly reducing the particle size of BCP. A maximum of 5 min of this pre-treatment, or even less, is sufficient to achieve a substantial particle size reduction and thereby increase the surface available for efficient mass and energy transfer during the extraction.

The effect of different cryogrinding exposure times on the content of TP and TMA in BCP extracts obtained under optimized MAE conditions (temperature 80 °C, extraction time 10 min, and SSR 1:80 g/mL) as well as on the antioxidant capacity of extracts is shown in Table 5.

Table 5. Content of TP and TMA and antioxidant capacity in BCP extracts obtained under MAE optimal conditions in combination with cryogrinding pre-treatment.

Cryogrinding Time (min)	TP (mg GAE/g dm)	TMA (mg C3GE/g dm)	FRAP (μmol TE/g dm)	DPPH (μmol TE/g dm)	ABTS (μmol TE/g dm)	ORAC (μmol TE/g dm)
	$p = 0.001$ *	$p = 0.980$	$p = 0.003$ *	$p = 0.046$ *	$p = 0.004$ *	$p = 0.006$ *
1	115.1 ± 0.2^{b}	22.0 ± 0.7^{a}	816.5 ± 6.7^{ab}	380.3 ± 3.3^{a}	631.5 ± 11.7^{a}	204.1 ± 4.1^{b}
3	103.4 ± 0.7^{a}	22.3 ± 0.9^{a}	822.8 ± 2.2^{ab}	395.3 ± 4.7^{ab}	634.6 ± 7.3^{a}	192.6 ± 3.4^{b}
5	100.7 ± 2.5^{a}	22.2 ± 0.9^{a}	945.2 ± 13.5^{c}	403.9 ± 4.4^{ab}	651.3 ± 16.0^{ab}	190.9 ± 1.1^{ab}
7	99.1 ± 0.2^{a}	22.4 ± 0.7^{a}	786.6 ± 22.4^{a}	401.3 ± 13.1^{ab}	688.4 ± 13.1^{bc}	189.6 ± 4.4^{ab}
9	104.2 ± 2.5^{a}	22.1 ± 0.4^{a}	881.5 ± 35.9^{bc}	411.7 ± 5.5^{b}	718.1 ± 14.6^{c}	177.1 ± 4.0^{a}

TP—total polyphenols; TMA—total monomeric anthocyanins; BCP—black chokeberry pomace; MAE—microwave-assisted extraction; GAE—gallic acid equivalents; C3GE—cyanidin-3-glucoside equivalents; TE—Trolox equivalents. Results are expressed as mean ± SD. * $p \leq 0.05$. Means with the same letter within a column are not significantly different at $p \leq 0.05$.

The duration of cryogrinding had a statistically significant impact on TP content as well as the antioxidant capacity measured by FRAP, DPPH, ABTS, and ORAC assays, whereas no statistically significant differences were observed in the TMA content between samples with respect to cryogrinding time. This can be explained by the fact that most anthocyanins are located in the fruit skin [34], and thus, their majority was likely released even during conventional grinding of BCP. This is supported by the similar TMA contents obtained from the conventionally ground sample subjected to MAE under optimal conditions and the average values of all cryogenically ground samples (24.0 vs. 22.2 mg C-3-GE/g dm).

The highest TP content (115.1 mg GAE/g dm) was recorded in the sample cryogenically ground for 1 min, while the lowest content (99.1 mg GAE/g dm) was observed after 7 min of cryogenic grinding. Since the 1 min cryoground sample exhibited a statistically significant higher TP content compared to all other samples, this pre-treatment could be considered the most effective for phenolic compound isolation. Compared to the conventionally ground sample, the superiority of cryogrinding was shown through a 17.6% increase in TP content (115.1 vs. 94.8 mg GAE/g dm).

Contrasting results were reported by Balbino et al. [35], who investigated cryogrinding as a pre-treatment to enhance the extraction of bioactive molecules from pumpkin seed cake. Their cryogrinding durations were 4, 8, and 12 min, and they observed increasing phenolic yields with longer grinding times, with the highest yield at 12 min. These differences may arise from the distinct structural properties of the materials. Pumpkin seed cake possesses a tougher, fibrous matrix that requires longer grinding to disrupt effectively and release entrapped phenolics [35]. In contrast, the softer structure of BCP likely allows rapid release of phenolics, rendering longer grinding times less effective.

Significantly higher antioxidant capacity determined by FRAP, DPPH, and ABTS assays were observed in samples ground for longer than 5 min. Specifically, the highest FRAP values were measured in samples ground for 5 and 9 min, the highest DPPH values in samples ground for 9 min, and the highest ABTS values in samples ground for 7 and 9 min. Prolonged cryogrinding likely contributed to the release of additional bioactive compounds in BCP, such as ascorbic acid or proanthocyanidins (condensed tannins), enhancing antioxidant potential. Polymeric proanthocyanidins often exhibit higher antioxidant activity than monomeric forms in assays such as DPPH and FRAP [36]. However, their contribution to TP content, as measured by the Folin–Ciocalteu assay, can be underestimated due to both structural and methodological limitations. The FC reagent reacts with a wide range of reducing substances and phenols, but its response can vary depending on phenolic structure. As a result, proanthocyanidins may react less predictably or intensely, and if they are poorly extracted, their actual contribution to TP content may not be fully captured by the assay [37]. However, it is notable that antioxidant capacity values measured by all three assays were comparable between the conventionally ground sample and those ground cryogenically (FRAP: 983.6 vs. 945.2 μmol TE/g dm; DPPH: 420.4 vs. 411.7 μmol TE/g dm; ABTS: 681.3 vs. 718.1 μmol TE/g dm). Balbino et al. [35] reported increased antioxidant capacity (measured by DPPH assay) in pumpkin seed cake samples subjected to the longest cryogenic grinding time (12 min) compared to the control. Rosa et al. [38] studied wheat bran extracts and found that antioxidant capacity (measured by ABTS assay) significantly increased with cryogenic grinding time from 3 to 20 min, with the highest value after 20 min. These results align with the trends observed in the present study.

On the other hand, the antioxidant capacity measured by ORAC assay showed the opposite trend, namely a decrease in antioxidant capacity with prolongation of cryogrinding time. A possible explanation for the observed decrease in antioxidant capacity measured by the ORAC assay, in line with the trend for TP content, may lie in the nature of the assay itself. The ORAC method is highly sensitive to free, low-molecular-weight phenolic compounds that act predominantly through hydrogen atom transfer mechanisms [32]. Mechanical grinding of the samples and, by extension, prolonged cryogrinding, although effective for cell disruption, may promote partial oxidation or polymerization of phenolics due to mechanical stress [39], leading to a reduction in extractable free phenolics. Consequently, while electron transfer-based assays such as FRAP, DPPH, and ABTS can still detect antioxidant capacity in polymerized or oxidized phenolics, the ORAC assay reflects a decrease due to the loss of these labile antioxidants.

In conclusion, cryogrinding proved to be an effective pre-treatment for the isolation of phenolic compounds from BCP, with only 1 min of grinding required to achieve maximum phenolic yield.

2.4. Phenolic Profile of BCP Extract Obtained Under Optimal MAE Conditions in Combination with Cryogrinding

The extract of BCP obtained under the optimal MAE conditions (80 °C/10 min/1:80 g/mL) with a cryogrinding pre-treatment of 1 min was further analyzed for individual phenolic compound composition using the UPLC-MS/MS method, and the results are presented in Table 6.

Table 6. Polyphenol composition in BCP extract obtained under optimal MAE conditions in combination with 1 min of cryogrinding pre-treatment.

Polyphenol	Precursor Ion (*m*/*z*)	Product Ion (*m*/*z*)	mg/g dm
	Anthocyanins		
Cyanidin-3-*O*-galactoside	449	287	20.67 ± 0.03
Cyanidin-3-*O*-glucoside	449	287	20.81 ± 0.01
Cyanidin-3-*O*-arabinoside	419	287	11.37 ± 0.06
Cyanidin-3-*O*-xyloside	419	287	11.31 ± 0.04
		Total	64.16
	Phenolic acids		
Isochlorogenic acid A	515	353	0.70 ± 0.01
Chlorogenic acid	353	191	0.80 ± 0.04
Neochlorogenic acid	353	191	0.91 ± 0.02
		Total	2.41
	Flavonols		
Quercetin-3-*O*-dihexoside	627	303	0.15 ± 0.01
Isorhamnetin-3-rutinoside	625	317	0.06 ± 0.01
Quercetin-3-*O*-rutinoside	611	303	0.91 ± 0.02
Isorhamnetin-pentosylhexoside	611	317	0.02 ± 0.04
Quercetin-3-*O*-vicianoside	597	434, 303	0.16 ± 0.01
Kaempferol-3-rutinoside	595	287	0.03 ± 0.05
Isorhamnetin-3-*O*-glucoside	479	317	0.02 ± 0.01
Quercetin-3-*O*-glucuronide	479	303	0.03 ± 0.02
Isorhamnetin-3-*O*-galactoside	479	317	0.02 ± 0.02
Quercetin-3-*O*-galactoside	465	303	1.54 ± 0.04
Quercetin-3-glucoside	465	303	1.61 ± 0.03
Kaempferol-3-glucoside	449	287	24.53 ± 0.03
		Total	29.08
	Flavan-3-ols and Procyanidins		
Epicatechin	291	139, 123	0.22 ± 0.05
Procyanidin B2	577	289	0.04 ± 0.04
		Total	0.26
Total UPLC-MS/MS polyphenols			95.91

BCP—black chokeberry pomace; MAE—microwave-assisted extraction. Results are expressed as mean ± SD.

In total, 21 phenolic compounds were identified, including 4 anthocyanins, 3 phenolic acids, 12 flavonols, 1 flavan-3-ol, and 1 procyanidin. Anthocyanins were the dominant group, accounting for 64.16 mg/g dm (66.9%). These findings align with Kaloudi et al. [2], who reported that anthocyanins constituted the largest proportion of TP in chokeberry skin (66%) compared to chokeberry flesh with seeds (10.8%) and whole dried berries (27%). They concluded that chokeberry skin contains up to 73% of the total fruit anthocyanins. Oszmiański and Lachowicz [4] also found that anthocyanins made up about 50% of the TP extracted from ground freeze-dried BCP, which is slightly lower than the proportion observed in this study. This difference may result from using different chokeberry cultivars or different pomace production as well as extraction methods.

Among the identified anthocyanins, cyanidin-3-glucoside and cyanidin-3-galactoside were the most abundant (20.81 and 20.67 mg/g dm, respectively), followed by cyanidin-3-arabinoside and cyanidin-3-xyloside, present in similar concentrations (11.37 and 11.31 mg/g dm). The distribution pattern only partly agrees with the results of Oszmiański and Lachowicz [4], who identified cyanidin-3-*O*-galactoside as the predominant anthocyanin in chokeberry pomace powder (7961.70 mg/100 g dm), followed by cyanidin-3-*O*-arabinoside (3116.02 mg/100 g dm). The same study also reported cyanidin-3-*O*-xyloside and cyanidin-3-*O*-glucoside at lower concentrations. Other studies have also confirmed cyanidin-3-*O*-arabinoside as the second most abundant anthocyanin in chokeberry products [2,40,41]. Conversely, the findings of the present study as well as the previous study of our research group [3] showed cyanidin-3-*O*-glucoside to be similar in concentration to cyanidin-3-*O*-galactoside, thereby being one of two major anthocyanins in BCP. The differences observed across studies may be attributed to cultivar variation and processing conditions. Notably, both this study and Elez Garofulić et al. [3] used the 'Nero' cultivar from the same source, explaining the similarity in anthocyanin profiles. However, Elez Garofulić et al. [3] reported a considerably lower total anthocyanin concentration (5.36 mg/g dm) than that determined in this study (64.16 mg/g dm). This discrepancy could be due to differences in pomace preparation; their pomace was produced using industrial pressing, which likely transferred a higher proportion of anthocyanins into the juice, leaving less in the pomace. Moreover, the pomace analyzed presently was freeze-dried, whereas a previous study used frozen pomace, which may further explain the higher concentrations obtained in this research.

When comparing the total anthocyanin content determined by spectrophotometric and UPLC-MS/MS analyses, a significant difference was noted; the UPLC-MS/MS method yielded three-fold higher anthocyanin levels. This can be explained by methodological differences. The pH differential method quantifies only monomeric anthocyanins by exploiting their reversible structural transformation under different pH conditions, while UPLC-MS/MS, being more sensitive and precise, detects and quantifies individual anthocyanin compounds, including minor or structurally complex derivatives that spectrophotometric methods typically fail to detect, and sums their concentrations. Similar findings were reported by Vagiri and Jensen [42], who observed that total anthocyanins in chokeberry pomace determined by chromatography were, on average, twice as high as TMA determined spectrophotometrically.

Flavonols represented the second-largest group (29.08 mg/g dm; 30.3% of identified phenolics). Kaempferol-3-glucoside was the most abundant (24.53 mg/g dm), with noticeable amounts of quercetin-3-glucoside and quercetin-3-galactoside, followed by quercetin-3-rutinoside, quercetin-3-vicianoside, and quercetin-3-dihexoside. Other flavonols were present at concentrations below 0.1 mg/g dm. Sójka et al. [5] reported flavonols as the least abundant group (0.86%) in chokeberry pomace, with quercetin-3-galactoside, quercetin-3-glucoside, and quercetin-3-*O*-rutinoside as the most abundant, aligning with the present results, although they did not detect kaempferol-3-glucoside. Oszmiański and Lachowicz [4] also reported a lower flavonol content (1.5%) with the same major quercetin derivatives.

Phenolic acids accounted for 2.5% of the identified phenolics (2.41 mg/g dm), with neochlorogenic and chlorogenic acids nearly equally represented, while isochlorogenic acid A was slightly less abundant. These results are in line with those of Sójka et al. [5], who found that phenolic acids constituted 1.5% of chokeberry pomace phenolics, with similar concentrations of chlorogenic and neochlorogenic acids. Oszmiański and Lachowicz [4] reported a higher proportion of phenolic acids (7.5%) but also similar relative concentrations of these acids.

Flavan-3-ols and procyanidins were the least abundant, representing only 0.2% and 0.04% of total identified phenolics, respectively. Although present in low concentrations, flavan-3-ols and procyanidins are recognized as potent antioxidants that significantly contribute to the overall antioxidant capacity of plant materials [43]. Oszmiański and Lachowicz [4] found that flavan-3-ols accounted for 1.3% of TP in chokeberry pomace, with epicatechin at higher levels (2.601 mg/g dry matter) and procyanidin B2 at lower levels.

Overall, the composition of individual phenolic compounds in chokeberry pomace reflects the influence of cultivar, pomace preparation, extraction optimization, and analytical method, emphasizing the importance of standardized processing and precise analytical techniques for accurate phenolic profiling.

3. Materials and Methods

3.1. Chemicals and Reagents

Distilled water was obtained by a Milli-Q water purification system (Millipore, Bedford, MA, USA). Ethanol (96%), iron (III) chloride hexahydrate, potassium chloride, and methanol were procured from Gram-mol (Zagreb, Croatia). Hydrochloric acid (37%) and glacial acetic acid were purchased from Carlo Erba Reagents (Val-de-Reuil, France), while the Folin–Ciocolateu reagent was obtained from Merck (Darmstandt, Germany). Sodium carbonate (anhydrous), sodium acetate trihydrate, and potassium persulfate were from Kemika (Zagreb, Croatia), and acetonitrile and fluorescein from Honeywell Riedel-de-Haën (Seelze, Germany). 2,4,6-tri(2-pyridyl)-s-triazine (TPTZ), 6-hydroxy-2,5,7,8-tetramethylchroman-2-carboxylic acid (Trolox), 2,2-diphenyl-1-(2,4,6-trinitrophenyl)hydrazyl (DPPH), 2,2′-azino-bis(3-ethylbenzothiazoline-6-sulfonic acid (ABTS), and 2,2′-Azobis(2-amidinopropane) dihydrochloride (AAPH) were from Sigma-Aldrich (St. Louis, MO, USA) and formic acid was from BDH Prolabo (Lutterworth, UK). Working standards for gallic acid, cyanindin-3-glucoside chloride, chlorogenic acid, quercetin-3-glucoside, kaempferol-3-*O* glucoside, and catechin were procured from Sigma-Aldrich.

3.2. Plant Material

Frozen black chokeberries (*A. melanocarpa* L.), harvested in 2023, were procured from Terra Food Ltd. (Koprivnica, Croatia). Upon receipt, the berries were immediately thawed at room temperature and processed into juice by pressing in a juice extractor (Hurom HU-100, Ver Vita Ltd., Zagreb, Croatia). The remaining BCP was collected and frozen at −80 °C/24 h before being freeze-dried at − 55 °C/48 h (Alpha 1-4 LSCPlus, Martin Christ Gefriertrocknungsanlagen GmbH, Osterode am Harz, Germany). The freeze-dried BCP was then packaged into polypropylene bags, vacuumed and hermetically sealed, and stored in a desiccator at room temperature for further analysis. The total solid content of the freeze-dried BCP was 97.79%, as determined using a moisture analyzer (Ohaus MB25, Parsippany, NJ, USA). Immediately prior to extraction, freeze-dried BCP used for the determination of optimal PLE, MAE, and UAE process conditions and for reflux extraction was ground using conventional grinding in an electric grinding machine (Waring WSG30, Sprzęt Laboratoryjny i Medyczny Labpartner KBS, Warszawa, Poland), while the freeze-dried BCP used for the determination of the effect of cryogrinding was ground using cryogenic liquid.

3.3. Extraction Procedure

3.3.1. PLE

The Dionex^{TM}ASE^{TM} 350 Accelerated Solvent Extractor (Thermo Fisher Scientific Inc., Sunnyvale, CA, USA) was used for the PLE of polyphenols and anthocyanins from freeze-dried BCP. Extractions were performed according to an experimental design (Table 1) that included variation of temperature (100, 125, and 150 °C), static extraction time (5 and 10 min), and SSR (1:40, 1:60, and 1:80 g/mL) using a 1% formic acid in 50% aqueous ethanol solution as extraction solvent. The procedure was performed in stainless steel cells (34 mL) fitted with two cellulose filters at the bottom and one at the top, and filled with a mixture of the appropriate amount of sample and diatomaceous earth at constant parameters: 10.34 MPa pressure, 3 extraction cycles, 30 sec nitrogen purge, and 30% volume flush while varying the parameters temperature and static extraction time as described above. The extracts obtained were filtered into a volumetric flask (50 mL) and filled with the extraction solvent. Afterwards, they were stored in plastic test tubes at +4 °C until analysis.

3.3.2. MAE

The MAE of BCP polyphenols and anthocyanins was performed with an Ethos Easy Reactor (Milestone, Sorisole, Italy). An appropriate amount of ground freeze-dried BCP was added to the extraction cell together with 20 mL of solvent (1% formic acid in 50% aqueous ethanol solution) and a magnetic stirrer. The sealed cell was then placed in the reactor, and the extractions were carried out under the experimental design data (Table 1), varying the temperature (40, 60, and 80 °C; constant throughout procedure), extraction time (5 and 10 min), and SSR (1:40, 1:60 and 1:80 g/mL), while the microwave power (400 W), preheating time (2 min for 40 °C, 4 min for 60 °C, and 6 min for 80 °C), post-extraction aeration and cooling (2 min), and stirring intensity (50%) were kept constant. The extracts obtained were filtered into volumetric flasks (25 mL), filled with a solvent, and stored in plastic test tubes at + 4 °C until analysis.

3.3.3. UAE

The ultrasound processor UP200Ht (Dr. Hielscher GmbH, Teltow, Germany; max. output power 200 W, ultrasound frequency 26 kHz) was used for the UAE of polyphenols and anthocyanins from BCP. An appropriate amount of ground freeze-dried BCP and 20 mL of solvent (1% formic acid in 50% aqueous ethanol solution) were mixed in a glass beaker (100 mL) and placed in an ice bath to ensure a maximum temperature of 30 °C during the extraction. The extractions were performed by immersing a titanium ultrasound probe (154 mm^2) 1 cm deep into the extraction mixture, following the experimental design (Table 1). The parameters varied were the ultrasound amplitude (50, 75, and 100%), the extraction time (5 and 10 min), and the SSR (1:40, 1:60, and 1:80 g/mL). During extraction, the temperature was monitored with an infrared thermometer. After extraction, the extracts were filtered into volumetric flasks (25 mL), filled with a solvent, and stored in plastic test tubes at + 4 °C until analysis.

3.3.4. Reflux

Conventional heat extraction with reflux was carried out by weighing 0.5 g of ground freeze-dried BCP into a round-bottom flask and mixing it with 40 mL of extraction solvent (1% formic acid in 50% aqueous ethanol solution). The extraction mixture was then heated under reflux for 30 min from the start of boiling. The resulting extract was filtered into a volumetric flask (50 mL), made up with a solvent, and stored in a plastic test tube at +4 °C until analysis.

3.4. Cryogrinding

For pre-treatment by cryogrinding, 15 g of freeze-dried BCP was weighed into the grinding cell (50 mL), together with a 10 cm stainless steel pin. The cell was then sealed with a stainless cap and ground in a laboratory-scale Spex 6875D Freezer/Mill (Metuchen, NJ, USA) under a stream of liquid nitrogen (Messer, Zagreb, Croatia). Prior to grinding, the sample was pre-cooled for 2 min. Cryogrinding was performed for 1, 3, 5, 7, and 9 min at a rate of 14 cycles per second (cps). The resulting powders were collected and stored in plastic test tubes at −18 °C until extraction.

3.5. Particle Size Measurement

Particle size analysis of the conventionally ground and cryoground BCP samples was performed using laser diffraction on the Malvern Mastersizer 2000 (Malvern Instruments, Worcestershire, UK). Data acquisition and processing were carried out using Mastersizer 2000 software (ver. 5.61). The measurements were conducted with the Scirocco dry dispersion unit, with approximately 3 g of powder loaded into the sample chamber of the dispersion unit. The measurements were performed in duplicate at a pressure of 1.5 bar and a feed rate of 100%. Values for the 10th, 50th, and 90th percentiles (d(0.1), d(0.5), and d(0.9); μm) of the cumulative size distribution of the powders were recorded under the following parameters: a refractive index of 1.5, absorption of 0.1, and obscuration level of 3%. Span values, representing the homogeneity of the particle size distribution, were calculated according to the equation

$$\text{Span} = (d(0.9) - d(0.1))/d(0.5) \quad (1)$$

3.6. Determination of Total Polyphenols

The TP content was determined by the spectrophotometric Folin–Ciocalteu method described by Elez Garofulić et al. [44]. For this purpose, an appropriately diluted extract (100 μL), the Folin–Ciocalteu reagent (200 μL), and distilled water (2 mL) were mixed, and 1 mL of sodium carbonate solution (20%, *w*/*v*) was added after 3 min. The reaction mixture was vortexed and then thermostated in a water bath (50 °C). After 25 min, the absorbance was measured at 765 nm against the blank containing 100 μL of extraction solvent instead of the extract and the same reagents. The results were expressed in mg GAE/g dm based on the gallic acid calibration curve.

3.7. Determination of Total Monomeric Anthocyanins

The spectrophotometric pH differential method [45] was used for the determination of TMA content. In brief, extract (1 mL) was mixed with 0.025 M potassium chloride buffer (pH 1.0) (4 mL) or with 0.4 M sodium acetate buffer (pH 4.5) (4 mL) and kept at room temperature. After 20 min, the absorbance was measured at 520 and 700 nm using distilled water as a blank. The TMA content was calculated as C3GE (mg/L) according to the equation

$$\text{TMA (mg/L)} = (A \cdot MW \cdot DF \cdot 10^3)/(\varepsilon \cdot l) \quad (2)$$

where

$A = (A_{520nm} - A_{700nm})_{pH=1.0} - (A_{520nm} - A_{700nm})_{pH=4.5}$;
MW = molecular weight of cyanidin-3-glucoside (449.2 g/mol);
DF = dilution factor;
10^3 = factor for conversion g to mg;
ε = molar absorption extinction coefficient of cyanidin-3-glucoside (26,900 L/mol cm);
l = cuvette thickness (1 cm).

The final results were expressed in mg of C3GE/g dm.

3.8. Phenolic Characterization by UPLC-MS/MS

The individual polyphenols in the BCP extract obtained under optimal MAE conditions combined with cryogrinding pre-treatment were determined by UPLC-MS/MS analysis. Prior to analysis, the extract was filtered through a 0.45 μm membrane filter. The analysis took place on an Agilent 1290 series RRLC instrument with a 6430 triple quadrupole mass spectrometer (Agilent, Santa Clara, CA, USA) and an ESI ion source, according to the method described by Elez Garofulić et al. [46]. ESI ionization was carried out in both positive and negative modes (*m*/*z* 100 to 1000) using nitrogen (99.999%, Messer, Zagreb, Croatia) as an induction cone and collision gas at a positive/negative capillary voltage of 4000/3500 V, a drying gas temperature of 300 °C, a flow rate of 11 L/h, and nebulizer pressure of 40 psi. A Zorbax Eclipse Plus C18 column (Agilent, 100 × 2.1 mm; 1.8 μm particle size) at 35 °C was used to separate the compounds, with an injection volume of 2.5 μL, a flow rate of 0.3 mL/min, and the following eluents: eluent A—0.1% formic acid (*v*/*v*) in water and eluent B—0.1% formic acid in acetonitrile (*v*/*v*). Agilent MassHunter workstation software (ver. B.04.01) was used for data acquisition and processing. All other method quality parameters are described in detail in the authors' study [46]. Identification and quantification were performed using calibration curves for the following standards: cyanindin-3-glucoside chloride, chlorogenic acid, quercetin-3-glucoside, kaempferol-3-*O*-glucoside, and catechin. For compounds lacking commercial reference standards, tentative identification was based on mass spectral data and literature reports of mass fragmentation patterns. Quantification involved using cyanindin-3-glucoside chloride calibration for cyanidin-3-*O*-galactoside, cyanidin-3-*O*-glucoside, cyanidin-3-*O*-arabinoside, and cyanidin-3-*O*-xyloside, chlorogenic acid calibration for neochlorogenic acid and isochlorogenic acid A, quercetin-3-glucoside calibration for quercetin-3-*O*-dihexoside, isorhamnetin-3-rutinoside, quercetin-3-*O*-rutinoside, isorhamnetin-pentosylhexoside, quercetin-3-*O*-vicianoside, isorhamnetin-3-*O*-glucoside, quercetin-3-*O*-glucuronide, isorhamnetin-3-*O*-galactoside, quercetin-3-*O*-galactoside and quercetin-3-glucoside, kaempferol-3-*O* glucoside calibration for kaempferol-3-rutinoside, and catechin calibration for epicatechin and procyanidin B2. All results were expressed in mg/g dm.

3.9. *Determination of Antioxidant Capacity*

3.9.1. Ferric Reducing Antioxidant Power (FRAP) Analysis

The FRAP analysis method by Benzie and Strain [47], with some modifications, was followed. Briefly, a mixture of distilled water (240 µL), appropriately diluted extract (80 µL), and FRAP reagent (a mixture of sodium acetate buffer (0.3 M, pH 3.6), 10 mM TPTZ solution in 40 mM hydrochloric acid, and 20 mM $FeCl_3 \cdot 6H_2O$ aqueous solution in the ratio 10:1:1) (2080 µL) was vortexed and thermostatted at 37 °C. After 5 min, the absorbance was measured at 593 nm against the blank containing 80 µL of extraction solvent instead of the extract and with the same reagents. The results were expressed as µmol of Trolox equivalents (TE)/g dm based on the Trolox calibration curve.

3.9.2. 2,2-Diphenyl-1-picrylhydrazyl Radical (DPPH) Scavenging Analysis

DPPH method was performed using the method of Braca et al. [48]: 0.75 mL of the appropriately diluted extract was combined with 1.5 mL of 0.2 mM DPPH solution and left in the dark at room temperature for 20 min. Afterwards, the absorbance was measured at 517 nm against the blank (100% methanol). The results were expressed as µmol TE/g dm based on the Trolox calibration curve.

3.9.3. 2,2-Azinobis(3-ethylbenzothiazoline-6-sulfonic Acid) (ABTS) Analysis

The ABTS antioxidant capacity was determined according to the modified method of Miller and Rice-Evans [49]. For this purpose, an aliquot of 160 µL of appropriately diluted extract and 2 mL of 1% $ABTS^{\bullet +}$ solution were mixed and left for 1 min to incubate. Then, the absorbance was measured at 734 nm against the blank (96% ethanol). The results were expressed as µmol TE/g dm according to the Trolox calibration curve.

3.9.4. Oxygen Radical Absorbance Capacity (ORAC) Analysis

The ORAC analysis method described by Elez Garofulić et al. [50] was applied using an automated 96-well microplate fluorescence plate reader (Clariostar, BMG LABTECH, Offenburg, Germany). The appropriately diluted extract was placed in a black plate containing a fluorescein solution (70.3 nM) and incubated at 37 °C for 30 min. After the first three cycles (representing the baseline signal), 240 mM AAPH was injected into each well to generate the peroxyl radical. Different Trolox dilutions (3.12–103.99 µM) were used in each plate as a reference standard, and fluorescence intensity (excitation at 485 nm and emission at 528 nm) was monitored every 90 s over a total measurement period of 120 min.

Data were analyzed by MARS 2.0 software, and the results were expressed as μmol TE/g dm according to the Trolox calibration curve.

3.10. Experimental Design and Statistical Analysis

The Statistica 12.0 software system (StatSoft, Inc., Tulsa, OK, USA) was used for the experimental design and statistical data processing. In the first part of the study, i.e., optimizing the conditions for PLE, MAE, and UAE, all experiments were designed as mixed full-factorial experimental designs, with three independent factors at two (extraction time) or three levels (extraction temperature or amplitude and SSR) for each extraction technique, while the dependent variables were the content of TP (mg GAE/g dm) and TMA (mg C3GE/g dm). The normality and homogeneity of the distribution of residual data were tested with the Shapiro–Wilks and Levene tests, where normally distributed and homogeneous data were analyzed with a parametric test of multifactorial analysis of variance (ANOVA), and the comparison of marginal means was performed with the post hoc Tukey test. Data that deviated from the normal distribution and were inhomogeneous were analyzed using the non-parametric Kruskal–Wallis test. The results of this statistical analysis are expressed as mean ± SE. Furthermore, to examine the possible separation of extracts according to extraction technique, a PCA was applied to the correlation matrix comprising 54 samples and data for the content of TP and TMA (108 data points in total), considering principal components with an eigenvalue > 1 and including variables with a communality value ≥ 0.5 in the analysis.

In the second part of the study, i.e., when comparing the efficiency of PLE, MAE, UAE and reflux extraction and the effect of cryogrinding time, the experiments were designed as a single-factor design in which one independent factor at four (PLE, MAE, UAE, and reflux) or five levels (1, 3, 5, 7, and 9 min) was observed with the same dependent variables as in the first part of the study. The influence of extraction technique or duration of cryogrinding on the efficiency of extraction of polyphenols and anthocyanins and antioxidant capacity (FRAP, DPPH, ABTS, ORAC; μmol TE/g dm) was analyzed using the same statistical methodology as described above. The results of this statistical analysis are expressed as mean ± SD. All statistical tests were based on the significance level $p \leq 0.05$.

4. Conclusions

In this study, the efficiency of PLE, MAE, and UAE for recovering TP and TMA from BCP was systematically evaluated and compared to conventional reflux extraction. The results clearly demonstrate that process parameters, particularly temperature, time, and SSR, play critical roles in maximizing bioactive compound yield and antioxidant capacity. Among the advanced techniques, PLE proved the most effective for maximizing TP content, benefiting from higher temperatures that enhance the extraction process, though it showed limitations in preserving thermolabile anthocyanins. In contrast, MAE emerged as the most balanced method overall, combining relatively high yields of both TP and TMA with notable antioxidant activity across multiple assays, likely due to efficient cell disruption at moderate temperatures and short processing times. UAE showed lower efficiency in TP extraction under the tested conditions, but maintained competitive TMA yields, especially considering its mild operational parameters. The integration of cryo-grinding as a pre-treatment notably enhanced TP recovery by reducing particle size and increasing surface area for mass transfer, with just 1 min of cryogrinding achieving abundant improvements. The phenolic profile analysis of optimized extracts confirmed that anthocyanins, particularly cyanidin-3-glucoside and cyanidin-3-galactoside, dominate the BCP matrix, complemented by flavonols, phenolic acids, and minor flavan-3-ols and procyanidins, altogether contributing to the strong antioxidant potential observed. When benchmarked against reflux extraction, advanced methods, especially MAE combined with cryogrinding, offer comparable or superior extraction efficiency while providing significant advantages in reduced processing time that aligns well with the principles of sustainable green extraction technologies. Importantly, this research highlights the substantial potential of BCP, a major by-product of juice production and typically underutilized, as a valuable source of health-promoting compounds. By optimizing modern green extraction technologies and integrating simple yet effective pre-treatments like cryogrinding, this study demonstrates a viable approach to valorizing agro-industrial residues, reducing food processing waste, and contributing to sustainable bio-economy models. Such strategies align with global efforts to promote resource efficiency, support circular economy principles, and develop value-added functional ingredients for the food and nutraceutical sectors.

Author Contributions: Conceptualization, M.R.; methodology, M.R.; formal analysis, M.Z., I.M. and S.P.; investigation, M.R., M.Z. and I.M.; data curation, M.R., M.Z., I.M., S.P. and I.E.G.; writing—original draft preparation, M.R. and I.E.G.; writing—review and editing, M.R., S.P., V.D.-U. and I.E.G.; visualization, M.R. and I.E.G.; supervision, M.R.; project administration, V.D.-U. and I.E.G.; funding acquisition, V.D.-U. and I.E.G. All authors have read and agreed to the published version of the manuscript.

Funding: This research was funded by the Croatian Science Foundation project, grant number IP-2022-10-5499.

Data Availability Statement: The original contributions presented in the study are included in the article material; further inquiries can be directed to the corresponding author.

Acknowledgments: The authors are grateful to Terra Food Ltd. for providing the chokeberries used in the experiment, and Zoran Herceg for his support in providing the laboratory equipment used in the experiments through the project "Equipping the Semi-Industrial Practice Laboratory for the De-velopment of New Food Technologies" (KK.01.1.1.02.0001) financed by the European Union.

Conflicts of Interest: The authors declare no conflicts of interest. The funders had no role in the design of the study; in the collection, analyses, or interpretation of data; in the writing of the manuscript; or in the decision to publish the results.

Abbreviations

The following abbreviations are used in this manuscript:

BCP	Black chokeberry pomace
PLE	Pressurized liquid extraction
MAE	Microwave-assisted extraction
UAE	Ultrasound-assisted extraction
TP	Total polyphenols
TMA	Total monomeric anthocyanins
SSR	Solid–solvent ratio
PCA	Principal component analysis
PC1, PC2	Principal components

References

1. Fraga, C.G.; Croft, K.D.; Kennedy, D.O.; Tomás-Barberán, F.A. The effects of polyphenols and other bioactives on human health. *Food Funct.* **2019**, *10*, 514–528. [CrossRef]
2. Kaloudi, T.; Tsimogiannis, D.; Oreopoulou, V. *Aronia melanocarpa*: Identification and exploitation of its phenolic components. *Molecules* **2022**, *27*, 4375. [CrossRef]

3. Elez Garofulić, I.; Repajić, M.; Zorić, Z.; Jurendić, T.; Dragović-Uzelac, V. Evaluation of microwave-and ultrasound-assisted extraction techniques for revalorization of black chokeberry (*Aronia melanocarpa*) fruit pomace anthocyanins. *Sustainability* **2023**, *15*, 7047. [CrossRef]
4. Oszmiański, J.; Lachowicz, S. Effect of the production of dried fruits and juice from chokeberry (*Aronia melanocarpa* L.) on the content and antioxidative activity of bioactive compounds. *Molecules* **2016**, *21*, 1098. [CrossRef]
5. Sójka, M.; Kołodziejczyk, K.; Milala, J. Polyphenolic and basic chemical composition of black chokeberry industrial by-products. *Ind. Crops Prod.* **2013**, *51*, 77–86. [CrossRef]
6. Tena, N.; Asuero, A.G. Up-to-date analysis of the extraction methods for anthocyanins: Principles of the techniques, optimization, technical progress, and industrial application. *Antioxidants* **2022**, *11*, 286. [CrossRef] [PubMed]
7. Elmas, E.; Şen, F.B.; Kublay, İ.Z.; Baş, Y.; Tüfekci, F.; Derman, H.; Bekdeşer, B.; Aşçı, Y.S.; Capanoglu, E.; Bener, M. Green Extraction of Antioxidants from Hazelnut By-products Using Microwave-Assisted Extraction, Ultrasound-Assisted Extraction, and Pressurized Liquid Extraction. *Food Bioprocess Technol.* **2025**, *18*, 5388–5406. [CrossRef]
8. Pereira, D.T.V.; Tarone, A.G.; Cazarin, C.B.B.; Barbero, G.F.; Martinez, J. Pressurized liquid extraction of bioactive compounds from grape marc. *J. Food Eng.* **2019**, *240*, 105–113. [CrossRef]
9. Machado, A.P.D.F.; Pasquel-Reátegui, J.L.; Barbero, G.F.; Martínez, J. Pressurized liquid extraction of bioactive compounds from blackberry (*Rubus fruticosus* L.) residues: A comparison with conventional methods. *Food Res. Int.* **2015**, *77*, 675–683. [CrossRef]
10. Taha, M.; Dimitrov, K.; Samaillie, J.; Caux, B.; Sahpaz, S.; Blanchemain, N.; West, C.; Rivière, C. Optimizing the Extraction of Bioactive Compounds (Polyphenols, Lipids, and Alpha-Tocopherol) from Almond Okara to Unlock Its Potential as Functional Food. *Foods* **2024**, *13*, 2828. [CrossRef]
11. Kraljić, K.; Škevin, D.; Čukelj Mustač, N.; Benković, M.; Drakula, S.; Balbino, S.; Mandura Jarić, A.; Mamilović, K.; Ramljak, I.; Ćurić, D. Influence of cryogenic grinding on the nutritional and antinutritional components of rapeseed cake. *Appl. Sci.* **2023**, *13*, 5841. [CrossRef]
12. Cvitković, D.; Lisica, P.; Zorić, Z.; Pedisić, S.; Repajić, M.; Dragović-Uzelac, V.; Balbino, S. The influence of cryogrinding on essential oil, phenolic compounds and pigments extraction from myrtle (*Myrtus communis* L.) leaves. *Processes* **2022**, *10*, 2716. [CrossRef]
13. Saxena, S.N.; Saxena, P.; Rathore, S.S.; Sharma, L.K.; Saxena, R.; Barnwal, P. Effect of cryogenic grinding on phenolic compounds and antioxidant properties of fenugreek (*Trigonella foenum-graecum* L.) seed extract. *J. Spices Aromat. Crops* **2016**, *25*, 73–78.
14. Jung, J.-Y.; Rhee, J.-K. Roasting and cryogenic grinding enhance the antioxidant property of sword beans (*Canavalia gladiata*). *J. Microbiol. Biotechnol.* **2020**, *30*, 1706. [CrossRef]
15. Woźniak, Ł.; Marszałek, K.; Skąpska, S.; Jędrzejczak, R. The application of supercritical carbon dioxide and ethanol for the extraction of phenolic compounds from chokeberry pomace. *Appl. Sci.* **2017**, *7*, 322. [CrossRef]
16. Simić, V.M.; Rajković, K.M.; Stojičević, S.S.; Veličković, D.T.; Nikolić, N.Č.; Lazić, M.L.; Karabegović, I.T. Optimization of microwave-assisted extraction of total polyphenolic compounds from chokeberries by response surface methodology and artificial neural network. *Sep. Purif. Technol.* **2016**, *160*, 89–97. [CrossRef]
17. D'Alessandro, L.G.; Dimitrov, K.; Vauchel, P.; Nikov, I. Kinetics of ultrasound assisted extraction of anthocyanins from *Aronia melanocarpa* (black chokeberry) wastes. *Chem. Eng. Res. Des.* **2014**, *92*, 1818–1826. [CrossRef]
18. Roda-Serrat, M.C.; Andrade, T.A.; Rindom, J.; Lund, P.B.; Norddahl, B.; Errico, M. Optimization of the recovery of anthocyanins from chokeberry juice pomace by homogenization in acidified water. *Waste Biomass Valorization* **2021**, *12*, 1815–1827. [CrossRef] [PubMed]
19. Dobroslavić, E.; Elez Garofulić, I.; Šeparović, J.; Zorić, Z.; Pedisić, S.; Dragović-Uzelac, V. Pressurized Liquid Extraction as a Novel Technique for the Isolation of *Laurus nobilis* L. Leaf Polyphenols. *Molecules* **2022**, *27*, 5099. [CrossRef]
20. Ju, Z.Y.; Howard, L.R. Effects of solvent and temperature on pressurized liquid extraction of anthocyanins and total phenolics from dried red grape skin. *J. Agric. Food Chem.* **2003**, *51*, 5207–5213. [CrossRef] [PubMed]
21. Wu, T.; Yan, J.; Liu, R.; Marcone, M.F.; Aisa, H.A.; Tsao, R. Optimization of microwave-assisted extraction of phenolics from potato and its downstream waste using orthogonal array design. *Food Chem.* **2012**, *133*, 1292–1298. [CrossRef]
22. Jokić, S.; Cvjetko, M.; Božić, Đ.; Fabek, S.; Toth, N.; Vorkapić-Furač, J.; Redovniković, I.R. Optimisation of microwave-assisted extraction of phenolic compounds from broccoli and its antioxidant activity. *Int. J. Food Sci. Technol.* **2012**, *47*, 2613–2619. [CrossRef]
23. Zhao, C.-N.; Zhang, J.-J.; Li, Y.; Meng, X.; Li, H.-B. Microwave-assisted extraction of phenolic compounds from *Melastoma sanguineum* fruit: Optimization and identification. *Molecules* **2018**, *23*, 2498. [CrossRef] [PubMed]
24. Wen, Y.; Chen, H.; Zhou, X.; Deng, Q.; Zhao, Y.; Zhao, C.; Gong, X. Optimization of the microwave-assisted extraction and antioxidant activities of anthocyanins from blackberry using a response surface methodology. *RSC Adv.* **2015**, *5*, 19686–19695. [CrossRef]

25. Pap, N.; Beszédes, S.; Pongrácz, E.; Myllykoski, L.; Gábor, M.; Gyimes, E.; Hodúr, C.; Keiski, R.L. Microwave-assisted extraction of anthocyanins from black currant marc. *Food Bioprocess Technol.* **2013**, *6*, 2666–2674. [CrossRef]
26. Borrás-Enríquez, A.J.; Reyes-Ventura, E.; Villanueva-Rodríguez, S.J.; Moreno-Vilet, L. Effect of ultrasound-assisted extraction parameters on total polyphenols and its antioxidant activity from mango residues (*Mangifera indica* L. var. Manililla). *Separations* **2021**, *8*, 94. [CrossRef]
27. González, M.J.A.; Carrera, C.; Barbero, G.F.; Palma, M. A comparison study between ultrasound–assisted and enzyme–assisted extraction of anthocyanins from blackcurrant (*Ribes nigrum* L.). *Food Chem. X* **2022**, *13*, 100192. [CrossRef]
28. He, B.; Zhang, L.-L.; Yue, X.-Y.; Liang, J.; Jiang, J.; Gao, X.-L.; Yue, P.-X. Optimization of ultrasound-assisted extraction of phenolic compounds and anthocyanins from blueberry (*Vaccinium ashei*) wine pomace. *Food Chem.* **2016**, *204*, 70–76. [CrossRef]
29. Osorio-Tobón, J.F. Recent advances and comparisons of conventional and alternative extraction techniques of phenolic compounds. *J. Food Sci. Technol.* **2020**, *57*, 4299–4315. [CrossRef] [PubMed]
30. Wu, X.; Beecher, G.R.; Holden, J.M.; Haytowitz, D.B.; Gebhardt, S.E.; Prior, R.L. Concentrations of anthocyanins in common foods in the United States and estimation of normal consumption. *J. Agric. Food Chem.* **2006**, *54*, 4069–4075. [CrossRef]
31. Christodoulou, M.C.; Orellana Palacios, J.C.; Hesami, G.; Jafarzadeh, S.; Lorenzo, J.M.; Domínguez, R.; Moreno, A.; Hadidi, M. Spectrophotometric methods for measurement of antioxidant activity in food and pharmaceuticals. *Antioxidants* **2022**, *11*, 2213. [CrossRef]
32. Prior, R.L.; Wu, X.; Schaich, K. Standardized methods for the determination of antioxidant capacity and phenolics in foods and dietary supplements. *J. Agric. Food Chem.* **2005**, *53*, 4290–4302. [CrossRef] [PubMed]
33. Wu, X.; Gu, L.; Prior, R.L.; McKay, S. Characterization of anthocyanins and proanthocyanidins in some cultivars of *Ribes*, *Aronia*, and *Sambucus* and their antioxidant capacity. *J. Agric. Food Chem.* **2004**, *52*, 7846–7856. [CrossRef]
34. Chaovanalikit, A.; Wrolstad, R.E. Total anthocyanins and total phenolics of fresh and processed cherries and their antioxidant properties. *J. Food Sci.* **2004**, *69*, FCT67–FCT72. [CrossRef]
35. Balbino, S.; Dorić, M.; Vidaković, S.; Kraljić, K.; Škevin, D.; Drakula, S.; Voučko, B.; Čukelj, N.; Obranović, M.; Ćurić, D. Application of cryogenic grinding pretreatment to enhance extractability of bioactive molecules from pumpkin seed cake. *J. Food Process Eng.* **2019**, *42*, e13300. [CrossRef]
36. Wei, S.-D.; Zhou, H.-C.; Lin, Y.-M. Antioxidant activities of fractions of polymeric procyanidins from stem bark of *Acacia confusa*. *Int. J. Mol. Sci.* **2011**, *12*, 1146–1160. [CrossRef]
37. Schofield, P.; Mbugua, D.M.; Pell, A.N. Analysis of condensed tannins: A review. *Anim. Feed Sci. Technol.* **2001**, *91*, 21–40. [CrossRef]
38. Rosa, N.N.; Barron, C.; Gaiani, C.; Dufour, C.; Micard, V. Ultra-fine grinding increases the antioxidant capacity of wheat bran. *J. Cereal Sci.* **2013**, *57*, 84–90. [CrossRef]
39. Sapna, I.; Kamaljit, M.; Priya, R.; Jayadeep, P.A. Milling and thermal treatment induced changes on phenolic components and antioxidant activities of pigmented rice flours. *J. Food Sci. Technol.* **2019**, *56*, 273–280. [CrossRef]
40. Lin, S.; Meng, X.; Tan, C.; Tong, Y.; Wan, M.; Wang, M.; Zhao, Y.; Deng, H.; Kong, Y.; Ma, Y. Composition and antioxidant activity of anthocyanins from *Aronia melanocarpa* extracted using an ultrasonic-microwave-assisted natural deep eutectic solvent extraction method. *Ultrason. Sonochem.* **2022**, *89*, 106102. [CrossRef] [PubMed]
41. Slimestad, R.; Torskangerpoll, K.; Nateland, H.S.; Johannessen, T.; Giske, N.H. Flavonoids from black chokeberries, *Aronia melanocarpa*. *J. Food Compos. Anal.* **2005**, *18*, 61–68. [CrossRef]
42. Vagiri, M.; Jensen, M. Influence of juice processing factors on quality of black chokeberry pomace as a future resource for colour extraction. *Food Chem.* **2017**, *217*, 409–417. [CrossRef]
43. Soobrattee, M.A.; Neergheen, V.S.; Luximon-Ramma, A.; Aruoma, O.I.; Bahorun, T. Phenolics as potential antioxidant therapeutic agents: Mechanism and actions. *Mutat. Res. Mol. Mech. Mutagen.* **2005**, *579*, 200–213. [CrossRef] [PubMed]
44. Elez Garofulić, I.; Repajić, M.; Cegledi, E.; Dobroslavić, E.; Dobrinčić, A.; Zorić, Z.; Pedisić, S.; Franković, T.; Breški, M.; Dragović-Uzelac, V. Green Approach to Enhance the Recovery of Polyphenols from Blackcurrant and Bilberry Leaves: Evaluation of Microwave-Assisted and Pressurized Liquid Extraction. *Molecules* **2024**, *29*, 1351. [CrossRef] [PubMed]
45. Lee, J.; Durst, R.W.; Wrolstad, R.E.; Barnes, K.W.; Eisele, T.; Giusti, M.M.; Haché, J.; Hofsommer, H.; Koswig, S.; Krueger, D.A.; et al. Determination of total monomeric anthocyanin pigment content of fruit juices, beverages, natural colorants, and wines by the pH differential method: Collaborative study. *J. AOAC Int.* **2005**, *88*, 1269–1278. [CrossRef]
46. Elez Garofulić, I.; Zorić, Z.; Pedisić, S.; Brnčić, M.; Dragović-Uzelac, V. UPLC-MS2 Profiling of Blackthorn Flower Polyphenols Isolated by Ultrasound-Assisted Extraction. *J. Food Sci.* **2018**, *83*, 2782–2789. [CrossRef]
47. Benzie, I.F.F.; Strain, J.J. The ferric reducing ability of plasma (FRAP) as a measure of "antioxidant power": The FRAP assay. *Anal. Biochem.* **1996**, *239*, 70–76. [CrossRef] [PubMed]
48. Braca, A.; De Tommasi, N.; Di Bari, L.; Pizza, C.; Politi, M.; Morelli, I. Antioxidant principles from *Bauhinia tarapotensis*. *J. Nat. Prod.* **2001**, *64*, 892–895. [CrossRef]

49. Miller, N.J.; Rice-Evans, C.A. Cinnamates and hydroxybenzoates in the diet: Antioxidant activity assessed using the ABTS•+ radical cation. *Br. Food J.* **1997**, *99*, 57–62. [CrossRef]
50. Elez Garofulić, I.; Kruk, V.; Martić, A.; Martić, I.; Zorić, Z.; Pedisić, S.; Dragović, S.; Dragović-Uzelac, V. Evaluation of polyphenolic profile and antioxidant activity of *Pistacia lentiscus* L. leaves and fruit extract obtained by optimized microwave-assisted extraction. *Foods* **2020**, *9*, 1556. [CrossRef]

molecules

MDPI

Article

Effect of Ultrasonic Frequencies on the Aqueous Extraction of Polyphenols, Chlorogenic Acid, and Quercetin from the Whole Fruit of Pitaya (*Hylocereus* spp.)

Wei-Ting Lian and Chun-Yao Yang *

Department of Food Science, Fu Jen Catholic University, New Taipei City 242062, Taiwan; wiltonlian2@gmail.com
* Correspondence: 133810@mail.fju.edu.tw; Tel.: +886-2-2905-3624

Abstract: The effect of ultrasonic frequencies of 40 kHz/300 W (U-40) and 120 kHz/300 W (U-120) on the aqueous extraction of bioactive compounds from dried whole-fruit powders (DPs) of red-peel/white-flesh (WFP) and red-peel/red-flesh (RFP) pitayas was investigated, and shaking at 120 rpm (S-120) was used for a comparison. The effects of temperature and the solid-to-liquid ratio on the extraction efficiencies of the total phenolic content (TPC) and ferric-reducing antioxidant power (FRAP) of WFP and RFP were evaluated. The impact of extraction time on the aqueous extraction of specific compounds, namely, chlorogenic acid (CGA) and quercetin, from WFP and RFP was assessed with extraction modes of U-40, U-120, and S-120. At 40 °C and a 1/20 (g DP/mL) solid-to-liquid ratio, the use of U-40 achieved higher TPC and FRAP values at 15 min than U-120 and S-120 for WFP. The use of U-40 and U-120 extracted higher amounts of free CGA and free quercetin from WFP and RFP at 15 and 60 min than S-120 but showed different extraction efficiencies for free CGA and free quercetin. This study demonstrates that different ultrasonic frequencies can be applied in the green extraction of target bioactive compounds for use in nutraceutical foods.

Keywords: ultrasonic frequency; green extraction; pitaya; polyphenols; chlorogenic acid; quercetin; antioxidant activity

1. Introduction

Pitaya (*Hylocereus* spp.), also called dragon fruit, is widely cultivated in subtropical and tropical regions, and it is rich in phytochemicals with pharmacological value, such as in antioxidant, anti-microbial, anti-diabetic, and anti-cancer applications [1–3]. Pitaya can be classified into several types according to the appearance of the peel and flesh, and the main varieties commonly consumed are *Hylocereus undatus* (red peel/white flesh), *Hylocereus polyrhizus* (red peel/red flesh), and *Hylocereus megalanthus* (yellow peel/white flesh) [1]. Pitaya contains vitamins, polysaccharides, flavonoids, betalains, and polyphenols, and it can also be used as a nutraceutical [4–6]. However, different varieties of pitaya show different contents of health-promoting phytochemicals, among which the main phenolic compounds are flavonoids and phenolic acid derivatives, including chlorogenic acid (CGA), caffeic acid, synaptic acid, gallic acid, ellagic acid, quercetin, isoquercetin, and rutin [7,8].

Tang et al. (2021) identified a total of 37 phenolic compounds, with CGA, caffeic acid, ferulic acid, and *p*-coumaric acid identified in red-fleshed pitaya and CGA, caffeic acid, ferulic acid, rutin, and isoquercitrin identified in white-fleshed pitaya [5]. Attar et al. (2022) reported that the total phenolic content and antioxidant capacity in red-fleshed pitaya were higher than those in white-fleshed ones, and some bioactive compounds, such as gallic acid, myricetin, caffeic acid, quercetin, and kaempferol, were identified in fruits

 https://doi.org/10.3390/molecules30153253

of both species [9]. Chlorogenic acid (CGA, 3-*O*-caffeoylquinic acid) has beneficial effects due to its anti-hypertensive, anti-diabetic, anti-neurodegenerative, and anti-lipidemic properties [10,11]. Quercetin (3,3,4,5,7-pentahydroxyflavone) shows antioxidative, anti-inflammatory, anti-proliferative, anti-carcinogenic, anti-diabetic, and anti-viral properties [12,13]. Hence, it would be valuable to extract CGA and quercetin from pitaya for use as additives in nutraceutical foods.

Conventional extraction methods, such as Soxhlet extraction, maceration, percolation, infusion, and decoction, are generally used when extracting bioactive compounds from plants and the fruit biomass by utilizing a solvent other than water [14]. The emerging extraction technologies frequently used are ultrasound-assisted extraction, microwave-assisted extraction, and supercritical fluid extraction, among which ultrasound, as an environmentally friendly technology, shows potential for the enhanced extraction of bioactive compounds from the internal structure of plant foods. This method has the advantages of reduced extraction time, energy use, CO emissions, and solvent while enhancing extraction yield, and water or other green solvents can be effectively employed [14]. Furthermore, the profiles of the phytochemicals extracted from plant foods are dependent on the extraction methods and the properties of the solvent and target compounds, the interaction between phenolic compounds and polysaccharides, and the affinity of the components at the interface [15]. Additionally, ultrasound with different ultrasonic frequencies could be applied to extract some specific compounds from plant foods with various impacts [16,17].

The phenolic compounds in pitaya are present in the pulp, seed, and peel [8,18]. Although the part of pitaya usually consumed is the pulp, the peel is also a rich source of polyphenols beneficial for health [18]. The processing of the edible and nonedible parts of pitaya is tremendously significant in providing high-nutritional products and reducing waste generation [19]. Using the green extraction process to extract bioactive compounds directly from the whole pitaya fruit (pulp and peel) would be a favorable route for the process's sustainability by reducing the waste peel.

The edible white-flesh/red-flesh parts of red-peel pitaya cultivated in Taiwan contain about 85.7%/85.8% water, 12.4%/12.3% total carbohydrates, 0.9%/1.1% crude protein, and 0.4%/0.2% crude fat, including vitamins, fatty acids, amino acids, betacyanins, and other phenolic compounds [20,21]. Due to the existence of abundant carbohydrates in pitaya, the impact of ultrasonic frequency on the extraction of phytochemicals could vary among pitaya varieties, thus influencing the extraction strategy for target compounds. In addition, Zhong et al. (2022) indicated that bound polyphenols were present in pitaya peel [18]. Past publications on the extraction of phytochemicals from pitaya have generally focused on extraction from its individual parts using either conventional or emerging extraction technologies [5,8,9,18], and the green extraction of bioactive compounds from the whole fruit using water has rarely been explored. Hence, the aim of this study was to investigate the effect of ultrasound at different frequencies on the aqueous extraction of bioactive compounds from the whole fruit (peel and flesh) of red-peel/white-flesh (WFP) and red-peel/red-flesh (RFP) pitayas containing abundant carbohydrates. The effects of extraction parameters on the total phenolic content (TPC) and antioxidant activity under various extraction modes were evaluated. Using water as a green solvent for the extraction of the valuable compounds CGA and quercetin from WFP and RFP, the impact of different ultrasonic frequencies (40 kHz and 120 kHz), shaking (120 rpm), and extraction time on the extraction efficiency of free CGA and free quercetin from WFP and RFP in the presence of abundant carbohydrates was assessed.

2. Results and Discussion

2.1. Effect of Extraction Mode

The main bioactive compounds and nutritional components in an extract generally vary across varieties, cultivated conditions, and the extraction method and conditions used [7,9,16]. Table 1 shows the effects of different extraction modes, i.e., ultrasound at 40 kHz/300 W (U-40), ultrasound at 120 kHz/300 W (U-120), and shaking at 120 rpm (S-120), on the extract yields, TPC (using gallic acid equivalent (GAE)), and FRAP (using ferrous sulfate heptahydrate equivalent (FSE)) values of the dried pitaya extracts (DPEs) obtained from WFP and RFP at 40 °C, 15 min, and a 1/20 solid-to-liquid (S/L) ratio (g/mL). It was found that the extract yields of WFP (67–69%) were higher than those of RFP (61–63%) for the tested extraction modes, and the use of U-120 produced a slightly higher amount of extract for both WFP and RFP. The results revealed that more hydrophilic or water-soluble compounds could be extracted from WFP than from RFP.

Table 1. Effect of extraction mode on the extract yield, total phenolic content (TPC), and ferric-reducing antioxidant power (FRAP) of the dried whole-fruit powders (DPs) from white-flesh (WFP) and red-flesh (RFP) pitayas.

Type of Pitaya	Extraction Mode	Extract Yield (% of DP)	TPC (mg GAE/g DPE)	FRAP (mg FSE/g DPE)
WFP	U-40	67.103 ± 1.057^{B}	3.399 ± 0.181^{A}	9.646 ± 0.407^{A}
	U-120	69.027 ± 0.952^{A}	3.050 ± 0.148^{B}	7.567 ± 0.320^{B}
	S-120	67.137 ± 0.649^{B}	3.105 ± 0.049^{B}	8.285 ± 0.611^{B}
RFP	U-40	61.963 ± 1.240^{a}	8.985 ± 0.180^{a}	29.903 ± 1.695^{a}
	U-120	63.173 ± 1.023^{a}	8.105 ± 0.119^{b}	28.415 ± 0.615^{a}
	S-120	62.173 ± 2.440^{a}	8.125 ± 0.290^{b}	29.339 ± 1.468^{a}

Conditions: 40 °C, 15 min extraction time, 1/20 S/L ratio (g/mL). U-40: ultrasound at 40 kHz and 300 W; U-120: ultrasound at 120 kHz and 300 W; S-120: shaking at 120 rpm; DPE: dried pitaya extract. Data are expressed as mean ± standard deviation (n = 3). Values with different superscript letters in the same column were found to be significantly different ($p < 0.05$) using Duncan's multiple range test.

As shown in Table 1, for WFP, the TPC and FRAP values obtained when using U-40 were higher than those obtained when using U-120 and S-120, with significant differences. For RFP, the TPC values followed the order of U-40 > S-120 ≥ U-120, with a significant difference between U-40 and S-120 but an insignificant difference between S-120 and U-120.

For WFP, the FRAP values followed the order of U-40 > S-120 ≥ U-120, which was consistent with the order of the TPC values; for RFP, the order of the FRAP values almost corresponded to that of the TPC values. This suggests that the antioxidant activity is mainly contributed from the polyphenols in both types of pitaya. The results demonstrate that the use of U-40 is more beneficial for enhancing the extraction efficiency of polyphenolic compounds from WFP and RFP but has different impacts.

When applying ultrasound in the extraction system, cavitation bubbles form in the liquid medium when the negative pressure in the rarefaction phase of an acoustic wave reaches a certain threshold value, which varies according to the combination of parameters, such as the hydrostatic pressure level, liquid temperature, and change rate of local pressure [22]. Nguyen et al. (2017) reported that the threshold of the mechanical effect was nearly equal to that of the cavitation effect at an ultrasonic frequency of less than 98 kHz but was significantly greater at a high frequency [23]. Hence, when using U-40 or U-120, the ultrasonic impact on the variations in the TPC values of WFP and RFP might be attributed to whether the mechanical or cavitation effect dominated. Bu and Alheshibri (2021) also reported that the bubbles generated at a lower frequency were larger than those generated at a higher frequency, with the released energy intensity approximately inversely proportional to the ultrasonic frequency [24]. The high energy intensity and cavitation effect achieved in the water medium using U-40 better facilitate the extraction of polyphenols than those achieved using U-120 and S-120; in the latter, hydromechanical stress is generated via shaking, promoting mixing and the interaction between the compounds and solvent to different extents [25].

2.2. Effect of Temperature

The effect of temperature (30 to 60 °C) on the aqueous extraction of bioactive compounds from WFP and RFP was explored using U-40 with a 1/20 (g/mL) S/L ratio and a 15 min extraction time. The extract yield, TPC, and FRAP results of the dried whole-fruit powders (DPs) are shown in Table 2. The extract yield of WFP (65.6–68.0%) was greater than that of RFP (60.7–62.0%) at various tested extraction temperatures, and higher extract yields could be obtained above 40 °C for both WFP and RFP.

Table 2. Effect of temperature on the extract yield, total phenolic content (TPC), and ferric-reducing antioxidant power (FRAP) of the dried whole-fruit powders (DPs) from white-flesh (WFP) or red-flesh (RFP) pitaya using U-40.

Type of Pitaya	Temperature (°C)	Extract Yield (% of DP)	TPC (mg GAE/g DP)	FRAP (mg FSE/g DP)
WFP	30	65.600 ± 0.782 B	2.213 ± 0.092 B	4.643 ± 0.369 B
	40	67.103 ± 1.057 AB	2.280 ± 0.111 AB	6.470 ± 0.179 A
	50	67.967 ± 0.627 A	2.346 ± 0.181 AB	6.349 ± 0.408 A
	60	67.897 ± 1.401 A	2.536 ± 0.166 A	6.824 ± 0.186 A
RFP	30	60.673 ± 0.943 a	5.355 ± 0.768 a	18.605 ± 3.034 a
	40	61.963 ± 1.240 a	5.568 ± 0.205 a	18.525 ± 1.017 a
	50	61.620 ± 1.846 a	5.613 ± 0.254 a	19.528 ± 0.369 a
	60	61.493 ± 1.557 a	5.425 ± 0.171 a	19.951 ± 1.296 a

Conditions: 15 min extraction time, 1/20 S/L (g/mL) ratio. U-40: ultrasound at 40 kHz and 300 W. Data are expressed as mean ± standard deviation ($n = 3$). Values with different superscript letters in the same column for the same type of pitaya were found to be significantly different ($p < 0.05$) using Duncan's multiple range test.

The TPC and FRAP values of WFP at an extraction temperature of 40 °C were higher than those at 30 °C, and the FRAP value of WFP was 6.470 mg FSE/g DP at 40 °C, showing a significant increase with the increase in temperature from 30 °C to 40 °C; in addition, the TPC and FRAP values of RFP slightly increased with the increase in temperature, but they showed insignificant differences. Additionally, the TPC and FRAP trends in both WFP and RFP were similar, indicating that the antioxidant activity could be related to the polyphenols in both types of pitaya; this also revealed that higher temperatures increase the penetration of the solvent and the solubility of the solutes and facilitate the transport of solutes from solid material [16,26].

2.3. Effect of Solid-to-Liquid Ratio

The S/L ratio can affect the extraction efficiency of bioactive compounds from plant materials, as well as the extract yield, antioxidant activity, and polyphenols in the extract. In this study, various S/L (1/10 to 1/40 g/mL) ratios were employed under conditions of 40 °C and a 15 min extraction time using U-40. For WFP and RFP, the extract yield, TPC, and FRAP values of the DPs are displayed in Table 3.

As shown in Table 3, the extract yield of WFP was 73.360% at a 1/40 S/L ratio, and it was significantly higher than that at 1/10 to 1/30 (g/mL) S/L ratios; at an S/L ratio of 1/20 (g/mL), the FRAP value of WFP was 6.470 mg FSE/g DP, which was slightly higher than that at other S/L ratios, but there was no significant difference. For RFP, better extract yields were obtained, 61.963% and 64.820% at 1/20 and 1/30 (g/mL) S/L ratios, respectively, as well as higher FRAP values, 19.407 and 18.525 mg FSE/g DP at 1/30 and 1/20 (g/mL) S/L ratios, respectively.

Table 3. Effect of solid-to-liquid ratio on the extract yield, total phenolic content (TPC), and ferric-reducing antioxidant power (FRAP) of the dried whole-fruit powders (DPs) obtained from white-flesh (WFP) or red-flesh (RFP) pitaya using U-40.

Type of Pitaya	Solid-to-Liquid Ratio (g/mL)	Extract Yield (% of DP)	TPC (mg GAE/g DP)	FRAP (mg FSE/g DP)
WFP	1/10	67.320 ± 0.864^{B}	2.125 ± 0.123^{B}	5.667 ± 0.265^{A}
	1/20	67.103 ± 1.057^{B}	2.280 ± 0.111^{AB}	6.470 ± 0.179^{A}
	1/30	68.570 ± 1.488^{B}	2.507 ± 0.190^{A}	6.078 ± 0.641^{A}
	1/40	73.360 ± 0.472^{A}	2.479 ± 0.109^{A}	6.064 ± 1.222^{A}
RFP	1/10	52.863 ± 2.125^{c}	4.552 ± 0.052^{b}	15.287 ± 0.707^{b}
	1/20	61.963 ± 1.240^{ab}	5.568 ± 0.205^{a}	18.525 ± 1.017^{a}
	1/30	64.820 ± 0.442^{a}	5.617 ± 0.306^{a}	19.407 ± 1.867^{a}
	1/40	59.480 ± 3.680^{b}	4.820 ± 0.453^{b}	17.071 ± 1.861^{ab}

Conditions: 40 °C, 15 min extraction time. U-40: ultrasound at 40 kHz and 300 W. Data are expressed as mean ± standard deviation (n = 3). Values with different superscript letters in the same column for the same pitaya were found to be significantly different ($p < 0.05$) using Duncan's multiple range test.

The TPC values of RFP or WFP obtained using different S/L ratios and U-40 are shown in Table 3. For RFP, the TPC values followed the order of TPC for 1/30 (S/L) ≥ TPC for 1/20 (S/L) > TPC for 1/40 (S/L) ≥ TPC for 1/10 (S/L), showing the same trend as the FRAP values. This phenomenon suggests that, at a higher S/L ratio (with less solvent), the cavitation effect could be somewhat impeded due to the higher viscosity of the solution [27].

Moreover, it should be noted that pitaya contains a large amount of carbohydrates [20,21]. When using water as a solvent in the extraction of polyphenols, more hydrophilic carbohydrates were extracted and accounted for the majority of the extract yield, and the total polyphenol content in the extract yield seemed to be rather small in comparison. It was found that, when using U-40 with the tested S/L ratios, although the extract yields of WFP were larger than those of RFP, the TPC values of WFP were smaller than those of RFP under the same extraction conditions. A reason for this may be that, under the same extraction conditions, the cavitation effect on the microstructures of WFP and RFP had different impacts on the transport of ingredients within the internal structure and on the interaction between the polyphenols and carbohydrates [16,18,26].

2.4. *Effect of Extraction Time*

The effects of extraction time (5, 10, 15, and 60 min) on the TPC and FRAP values of the DPs obtained from WFP and RFP were evaluated with different extraction modes (U-40, U-120, and S-120) at 40 °C and a 1/20 (g/mL) S/L ratio. The TPC values of WFP are shown in Figure 1.

As shown in Figure 1, when using U-40, the TPC value at 10 min (2.489 mg GAE/g DP) was significantly higher than that at 60 min (2.020 mg GAE/g DP) and was slightly higher than that at 5 and 15 min with insignificant differences, revealing that using U-40 could lead to rather stable TPC values for 5 to 15 min with a decrease until 60 min. Although the TPC value at 15 min was slightly higher when using U-40 than when using U-120 and S-120 with insignificant differences, that at 10 min was significantly higher when using U-40 than when using U-120 and S-120, thus showing that U-40 could release a larger energy intensity than U-120 and S-120. However, the TPC value at 60 min was still higher when using U-40 than when using U-120 with a significant difference. This indicates that, with a longer extraction time, the phenolic compounds slightly degraded to various extents when using U-40 and U-120.

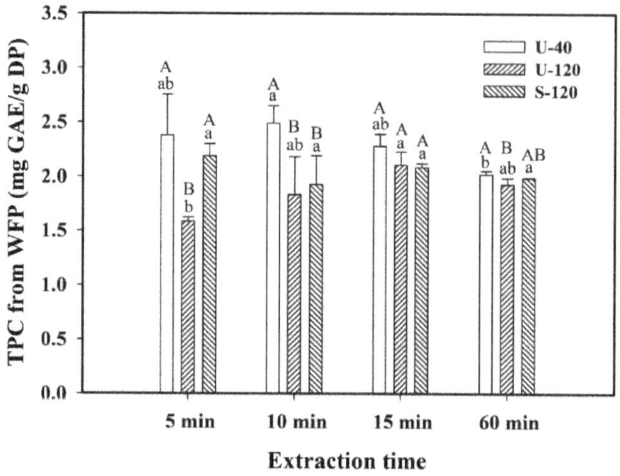

Figure 1. Variations in the total phenolic content (TPC) extracted from white-flesh (WFP) pitaya at different extraction times. Conditions: 40 °C, 1/20 S/L ratio (g/mL). U-40: ultrasound at 40 kHz and 300 W; U-120: ultrasound at 120 kHz and 300 W; S-120: shaking at 120 rpm. Data are expressed as mean ± standard deviation ($n = 3$). Values with different uppercase letters for the same extraction time or values with different lowercase letters for the same extraction mode were found to be significantly different ($p < 0.05$) using Duncan's multiple range test.

Moreover, when using U-120, the TPC value at 15 min (2.106 mg GAE/g DP) was significantly higher than that at 5 min (1.584 mg GAE/g DP) but was only slightly higher than that at 10 and 60 min with insignificant differences; this suggests that the possible degradation of phenolic compounds could be compensated within the first 15 min of extraction due to the enhanced extraction of U-120 at a lower energy intensity. Wang et al. (2020) investigated the degradation behavior of polyphenols in a model aqueous extraction system under ultrasound and reported that the hydroxyl radical concentration increased linearly with the sonication time, leading to the degradation of caffeic acid on a time scale of 25 min; thus, free radical scavengers could minimize the degradation of polyphenols by inhibiting hydroxyl radicals [26]. This revealed that the extraction efficiency of polyphenols was influenced to various extents by the combined effects of local hot spots generated by ultrasound, cavitation caused by different ultrasonic frequencies, and the interaction between extracted compounds [18,28].

The TPC values of RFP are displayed in Figure 2. For RFP, the TPC values of the DP at 15, 60, 5, and 10 min using U-40 were 5.568, 5.396, 5.200, and 5.109 mg GAE/g DP, respectively, with insignificant differences from the 5 to 60 min extraction times. When using U-120, the TPC values at 60, 10, and 15 min were 5.170, 5.139, and 5.119 mg GAE/g DP, respectively, with insignificant differences for these extraction times. The results revealed that, for RFP, the degradation levels of polyphenols were not significant after 15 min of extraction when using either U-40 or U-120, implying that their degradation could be compensated by sonication, which enhanced the release of phenolic compounds. However, although after 15 min of extraction, the TPC values of RFP were slightly higher when using U-40 than when using U-120 and S-120 with insignificant differences, using U-40 still had the benefits of obtaining extracts from RFP with stable and higher polyphenol contents.

Table 4 displays the effect of extraction time on the FRAP values of the extracts from WFP and RFP using different extraction modes at 40 °C and a 1/20 S/L ratio (g/mL). The results show that the FRAP values of the extracts obtained from RFP were larger than those obtained from WFP for each extraction mode and time. The FRAP results of WFP and

RFP could be related to the TPC values of WFP (Figure 1) and RFP (Figure 2), respectively, among which the TPC values of RFP were greater than those of WFP for the same extraction mode at each extraction time, demonstrating that polyphenols mainly contributed to the antioxidant activity of the extract.

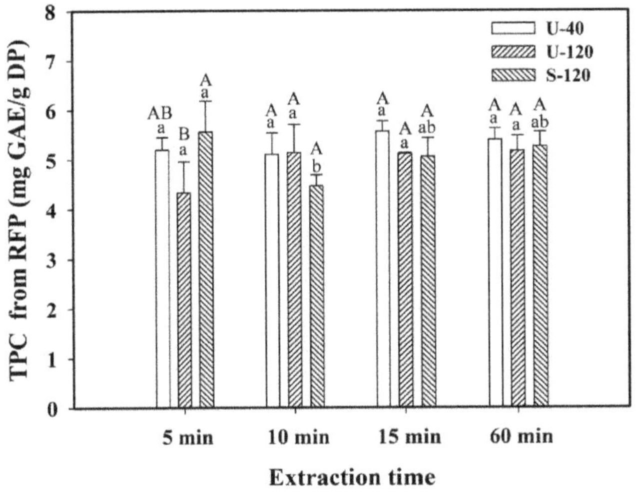

Figure 2. Variations in the total phenolic content (TPC) in red-flesh (RFP) pitaya at different extraction times. Conditions: 40 °C, 1/20 S/L ratio (g/mL). U-40: ultrasound at 40 kHz and 300 W; U-120: ultrasound at 120 kHz and 300 W; S-120: shaking at 120 rpm. Data are expressed as mean ± standard deviation (n = 3). Values with different uppercase letters for the same extraction time or values with different lowercase letters for the same extraction mode were found to be significantly different ($p < 0.05$) using Duncan's multiple range test.

Table 4. Effect of extraction time on ferric-reducing antioxidant power (FRAP) of the dried whole-fruit powders (DPs) obtained from white-flesh (WFP) or red-flesh (RFP) pitaya for different extraction modes.

Type of Pitaya	Extraction Mode	FRAP (mg FSE/g DP) and Extraction Time			
		5 min	10 min	15 min	60 min
WFP	U-40	6.879 ± 1.086 ^{aA}	7.291 ± 1.762 ^{aA}	6.470 ± 0.179 ^{aA}	5.443 ± 0.192 ^{aA}
	U-120	6.745 ± 0.404 ^{aA}	6.023 ± 1.090 ^{abA}	5.222 ± 0.184 ^{bB}	5.056 ± 0.157 ^{bA}
	S-120	6.457 ± 1.645 ^{aA}	5.837 ± 1.085 ^{aA}	5.560 ± 0.360 ^{aB}	5.343 ± 0.591 ^{aA}
RFP	U-40	18.834 ± 1.468 ^{aA}	17.764 ± 3.166 ^{aA}	18.525 ± 1.017 ^{aA}	17.443 ± 1.406 ^{aA}
	U-120	18.483 ± 1.442 ^{aA}	17.750 ± 0.575 ^{aA}	17.949 ± 0.329 ^{aA}	17.385 ± 1.371 ^{aA}
	S-120	18.642 ± 1.355 ^{aA}	19.213 ± 1.003 ^{aA}	18.256 ± 1.463 ^{aA}	17.897 ± 0.248 ^{aA}

Conditions: 40 °C, 1/20 S/L ratio (g/mL). U-40: ultrasound at 40 kHz and 300 W; U-120: ultrasound at 120 kHz and 300 W; S-120: shaking at 120 rpm. Data are expressed as mean ± standard deviation (n = 3). For the same type of pitaya, values with different uppercase letters for the same extraction time or values with different lowercase letters for the same extraction mode were found to be significantly different ($p < 0.05$) using Duncan's multiple range test.

When using U-40, the FRAP value of WFP was slightly higher at 10 min than at 5, 15, and 60 min with insignificant differences, revealing a slight decrease after 10 min. In addition, with a 15 min extraction time, the FRAP value obtained using U-40 was larger than that obtained using U-120 and S-120 with significant differences, but, with other extraction times, it was slightly higher than that obtained using U-120 and S-120 with insignificant differences, showing that U-40 can achieve relatively higher and stable FRAP values for WFP. For RFP, the FRAP values obtained at each extraction time using U-40,

U-120, and S-120 showed insignificant differences; however, they showed a slight decrease at 60 min. This suggests that, for RFP, an extraction time of 15 min is better for achieving an aqueous extract with more stable antioxidant activity.

2.5. Aqueous Extraction of Chlorogenic Acid and Quercetin

Chlorogenic acid (CGA) and quercetin are useful phytochemicals beneficial for human health, and their extraction from pitaya is valuable. The effects of extraction time and mode on the aqueous extraction of free CGA and free quercetin were assessed at 40 °C and a 1/20 S/L ratio (g/mL). Pitaya contains an abundance of carbohydrates [20]; thus, when using water as a green solvent to extract the free CGA from pitaya, a large amount of carbohydrates could be simultaneously extracted and interact with CGA during extraction such that the interaction between CGA and carbohydrates at various extraction durations and modes could occur to different extents [16,17]. The free CGA and free quercetin results are shown in Tables 5 and 6, respectively.

Table 5. Effect of extraction time on the free chlorogenic acid (CGA) content in the dried whole-fruit powders (DPs) obtained from white-flesh (WFP) or red-flesh (RFP) pitaya for different extraction modes.

Type of Pitaya	Extraction Mode	Free CGA (μg/g DP) and Extraction Time			
		5 min	10 min	15 min	60 min
WFP	U-40	67.78 ± 4.11 $^{b\,B}$	75.92 ± 7.53 $^{ab\,A}$	101.68 ± 31.12 $^{a\,A}$	71.58 ± 5.93 $^{ab\,A}$
	U-120	72.00 ± 3.56 $^{a\,AB}$	74.13 ± 8.28 $^{a\,A}$	76.15 ± 5.89 $^{a\,A}$	71.44 ± 5.67 $^{a\,A}$
	S-120	75.25 ± 1.40 $^{a\,A}$	79.55 ± 2.10 $^{a\,A}$	67.85 ± 4.18 $^{b\,A}$	65.25 ± 3.90 $^{b\,A}$
RFP	U-40	8.68 ± 0.96 $^{b\,A}$	11.07 ± 1.08 $^{a\,A}$	10.25 ± 1.71 $^{ab\,A}$	10.81 ± 0.42 $^{ab\,B}$
	U-120	(n.d.) $^{c\,B}$	10.82 ± 0.87 $^{ab\,A}$	9.63 ± 2.34 $^{b\,A}$	12.87 ± 0.18 $^{a\,A}$
	S-120	(n.d.) $^{c\,B}$	11.05 ± 0.25 $^{a\,A}$	8.93 ± 0.82 $^{b\,A}$	9.36 ± 0.50 $^{b\,C}$

Conditions: 40 °C, 1/20 S/L ratio (g/mL). U-40: ultrasound at 40 kHz and 300 W; U-120: ultrasound at 120 kHz and 300 W; S-120: shaking at 120 rpm. Data are expressed as mean ± standard deviation (n = 3). (n.d.): not detected. Values with different uppercase letters for the same extraction time and type of pitaya or values with different lowercase letters for the same extraction mode and type of pitaya were found to be significantly different ($p < 0.05$) using Duncan's multiple range test.

Table 6. Effect of extraction time on the free quercetin content in the dried whole-fruit powders (DPs) obtained from white-flesh (WFP) or red-flesh (RFP) pitaya using different extraction modes.

Type of Pitaya	Extraction Mode	Free Quercetin (μg/g DP) and Extraction Time			
		5 min	10 min	15 min	60 min
WFP	U-40	4.92 ± 0.43 $^{b\,A}$	5.51 ± 0.53 $^{ab\,AB}$	8.69 ± 3.40 $^{a\,A}$	7.17 ± 0.27 $^{ab\,AB}$
	U-120	5.55 ± 0.58 $^{b\,A}$	4.84 ± 0.90 $^{b\,B}$	5.82 ± 0.68 $^{b\,A}$	7.67 ± 0.45 $^{a\,A}$
	S-120	4.93 ± 0.45 $^{c\,A}$	6.24 ± 0.13 $^{a\,A}$	5.58 ± 0.44 $^{b\,A}$	6.79 ± 0.21 $^{a\,B}$
RFP	U-40	(n.d.) ^{c}	(n.d.) ^{c}	2.26 ± 0.57 $^{b\,A}$	3.25 ± 0.37 $^{a\,A}$
	U-120	(n.d.) ^{c}	(n.d.) ^{c}	2.41 ± 0.33 $^{b\,A}$	3.00 ± 0.40 $^{a\,AB}$
	S-120	(n.d.) ^{b}	(n.d.) ^{b}	(n.d.) $^{b\,B}$	2.42 ± 0.40 $^{a\,B}$

Conditions: 40 °C, 1/20 solid-to-liquid ratio (g/mL). U-40: ultrasound at 40 kHz and 300 W; U-120: ultrasound at 120 kHz and 300 W; S-120: shaking at 120 rpm. Data are expressed as mean ± standard deviation (n = 3). (n.d.): not detected. Values with different uppercase letters for the same extraction time and type of pitaya or values with different lowercase letters for the same extraction mode and type of pitaya were found to be significantly different ($p < 0.05$) using Duncan's multiple range test.

As displayed in Table 5, the free CGA contents in the DPs of WFP were larger than those in RFP at each extraction time and under each extraction mode. When using U-40, the free CGA content in WFP at 15 min was higher than that at 5 min with a significant difference but was slightly higher than that at 10 and 60 min with insignificant differences; furthermore, the CGA content at 15 min was found to have a larger standard deviation, which was speculated to result from the different degrees of interaction between CGA and carbohydrates during the extraction of the independent samples. Although the free CGA content extracted using U-40 slightly decreased at 60 min, the difference was insignificant.

When using S-120, the free CGA contents in WFP were significantly higher at 5 and 10 min than at 15 and 60 min. This phenomenon revealed that, after using S-120 for 15 min, severe degradation of the free CGA and possible interactions between CGA and carbohydrates occurred; this degradation was speculated to have been caused by the action of hydromechanical stress induced by shaking [29]. The results also revealed that the free CGA contents extracted at 15 and 60 min were higher when using U-40 and U-120 than when using S-120, showing that using ultrasound with a 15 min extraction time could provide a better CGA content in the extract than using shaking. Additionally, when using U-120, the free CGA content was slightly higher at 15 min than at the other extraction times but with insignificant differences, demonstrating that the free CGA in WFP was more stable throughout the extraction duration when using U-120 than when using U-40, as well as revealing that U-40 and U-120 have different impacts on the extraction of free CGA.

For RFP, free CGA was not detected at 5 min when using U-120 or S-120. However, when using S-120, the free CGA content was significantly higher at 10 min than at 60 and 15 min, indicating the severe degradation of free CGA after 15 min. Additionally, when using U-120, the free CGA content in RFP at 60 min was higher than that at 15 min with a significant difference, but it was slightly higher than that at 10 min with an insignificant difference. When using U-40, only a slight decrease was observed in the free CGA content in RFP from 10 to 60 min, implying a lower level of degradation of free CGA. These phenomena in RFP could be attributed to either the combined effects of free CGA interacting with carbohydrates and various levels of cavitation with local hot spots generated by different ultrasonic frequencies or to hydromechanical stress with fluid shear when using shaking [16,26,29].

Table 6 shows the effect of extraction time on the free quercetin content in the DPs of WFP and RFP when using different extraction modes. It was found that, under the same extraction mode and conditions, the free quercetin contents in WFP were larger than those in RFP. When using U-120, the free quercetin content in WFP was significantly higher at 60 min than at 5, 10, and 15 min, with no observable degradation of free quercetin after 15 min. Conversely, when using U-40, although having a larger standard deviation for a similar reason to the free CGA, the free quercetin content at 15 min was higher than that at 5 min with a significant difference, but it was slightly higher than that at 10 and 60 min with insignificant differences, showing possible degradation with a long extraction time (60 min) when using U-40 for WFP. This was speculated to result from the generation of free radicals in the aqueous solution and the thermal effects of the generation of local hot spots during sonication, leading to the degradation of polyphenols when using U-40 for ultrasonic irradiation for a long time (60 min) [29]. When using S-120, the significant increase in the free quercetin content from 15 to 60 min revealed a distinct effect of hydromechanical stress with fluid shear by shaking on the aqueous extraction of free quercetin from WFP.

For RFP, the free quercetin contents using both U-40 and U-120 significantly increased with an increasing time from 15 to 60 min, but with a slight variation in enhancement, showing that different ultrasonic frequencies had different impacts on the extraction efficiencies of free quercetin. In addition, at 5 and 10 min, free quercetin was not detected in RFP when using U-40 or U-120; moreover, when using S-120, no free quercetin could be detected at 5, 10, or 15 min. Quercetin is a more hydrophobic compound, with its solubility being 0.0000653 of mole fraction at 314.1 K in water [30]. Hence, using water to extract free quercetin from the internal structure of plant foods is limited not only by its solubility in water but also by the difficulty of its transport into the bulk solution, especially in the presence of abundant carbohydrates that could interact with polyphenols in pitaya. However, severe destruction of the microstructure and cell walls under ultrasonic irradiation is beneficial for enhancing the extraction efficiency of free quercetin, even when using water as a solvent. The results show that, for RFP, the highest free quercetin content was obtained with a 60 min extraction time, and the free quercetin content in RFP was significantly higher when using U-40 than when using S-120; this demonstrates that the extraction of hydrophobic free quercetin was effectively enhanced when using ultrasound due to the significant cavitation effect induced by sonication, especially when using the high-intensity and low-frequency ultrasound of U-40. The results indicate that ultrasound at 40 kHz and 120 kHz can promote the extraction efficiency of free quercetin better than shaking.

3. Materials and Methods

3.1. Materials

The chemical reagents 2,4,6-tris(2-pyridyl)-s-triazine, gallic acid, rutin, and ferric chloride were purchased from Merck KGaA (Darmstadt, Germany). Folin–Ciocalteu phenol reagent was purchased from Sigma-Aldrich Co. (St. Louis, MO, USA). Chlorogenic acid (≥95% purity) purchased from Merck KGaA (Darmstadt, Germany) and quercetin (97% purity) purchased from Thermo Fisher Scientific (Waltham, MA, USA) were used as standards in analyses. Other chemicals were purchased from Bionovas biotechnology Co. (Toronto, ON, Canada), J.T. Baker Chemical Company (Phillipsburg, NJ, USA), Merck KGaA (Darmstadt, Germany), and Taiwan Sugar Corporation (Tainan City, Taiwan).

For the experiments, red-peel/white-flesh pitaya (WFP) was purchased in July 2023 and red-peel/red-flesh pitaya (RFP) was purchased in December 2022 from an orchard in Erlin, ChangHua County, Taiwan (R.O.C.). After cleaning the surface of the pitaya with water, the scale leaves on the skin were cut off, and the head and stalk were removed. Then, the fruit was sliced into pieces for freeze-drying. The dried materials were ground and subsequently screened using a 40-mesh sieve to obtain dried whole-fruit powders (DPs), which were preserved in a drying cabinet until use. In addition, the moisture contents in WFP and RFP were determined to be 4.17 ± 0.04 % and 4.47 ± 0.01 % (n = 3) according to the method of Hsu and Yang (2024) [31].

3.2. Ultrasound-Assisted Extraction of Bioactive Compounds

The aqueous extraction of bioactive compounds from the DPs was carried out in a constant-temperature bath system, which was equipped with ultrasound at 300 W and a 40 kHz (or 120 kHz) frequency (LEO-3002S or LEO-3002H, LEO Ultrasonic Co., New Taipei City, Taiwan). Then, 1 g of the DP and definite volumes of de-ionized water were introduced into a 250 mL flask in a selected solid-to-liquid (S/L) ratio (grams of DP per mL of water, g/mL) at the set temperature. Extraction was then performed under ultrasound (40 kHz/300 W (U-40) or 120 kHz/300 W (U-120)); extraction was also conducted without ultrasound but with shaking at 120 rpm (S-120) (reciprocal shaking bath, Model B602D, Firstek, Taipei, Taiwan) for a comparison. Upon the completion of extraction, the liquid was separated from the solid–liquid mixture via centrifugation at 6000 rpm for 15 min. Afterward, freeze-drying was used to remove the water from the liquid solution, and dried pitaya extract (DPE) was obtained. The yield of extract from the DPs was estimated using Equation (1):

$$\text{Extract yield (\% of DP)} = (\text{weight of DPE}/\text{weight of DP}) \times 100\% \quad (1)$$

3.3. Determination of Total Phenolic Content

The TPC in the DPE was determined through the method of Kujala et al. (2000) with modifications using the gallic acid equivalent (denoted as GAE) [32]. A liquid sample of the DPE was prepared in a concentration of 5 mg/mL by re-dissolving it in water. The liquid sample (250 μL) was used to react with Folin–Ciocalteu reagent (1N, 250 μL) at 25 °C for 5 min. Afterward, 500 μL of 20% Na_2CO_3 was introduced into the mixture and reacted for 8 min. A spectrophotometer (Hitachi, Ratio Beam Spectrophotometer U-5100, Tokyo, Japan) was used to measure the absorbance of the above mixture at 730 nm. The TPC is expressed as mg GAE/g DPE or mg GAE/g DP, obtained by multiplying mg GAE/g DPE with g DPE/g DP.

3.4. Determination of Ferric-Reducing Antioxidant Power

By slightly modifying the method of Madane et al. (2020), the antioxidant activity of the DPE was evaluated according to the FRAP using the ferrous sulfate heptahydrate equivalent (denoted as FSE) [33]. A reagent for analyzing the FRAP was prepared and stored at 37 °C. Then, 15 μL of the liquid sample of the DPE (5 mg/mL in water), 285 μL of FRAP reagent, and 500 μL of de-ionized water were mixed to react at 37 °C for 4 min. The absorbance of the mixture at 593 nm was measured to determine the FRAP of the DPE sample. The FRAP is expressed as mg FSE/g DPE or mg FSE/g DP.

3.5. HPLC Analysis of CGA and Quercetin

An analysis of the CGA and quercetin contents in the DPE was performed with high-performance liquid chromatography (HPLC) following the method of Chen and Yang (2020) with a slight modification [34]. In the HPLC system, a Mightysil RP-18 GP column (5 μm, 250 mm × 4.6 mm, Kanto Chemical Co., Tokyo, Japan) and a UV–VIS detector (Hitachi Chromaster 5420 UV-VIS detector, Hitachi, Ltd., Tokyo, Japan) at a wavelength of 350 nm were used. A liquid sample of the DPE was prepared by re-dissolving it in 20% aqueous ethanol for an HPLC analysis at 30 °C. The mobile phase was solvent A (acetonitrile/0.1% trifluoroacetic acid at 10/90 (v/v)) and solvent B (acetonitrile), and the flow rate was 1.0 mL/min. The gradients of the mobile phase were set for solvent A as follows: 100 to 88.9% at 0 to 10 min, 88.9 to 66.7% at 10 to 35 min, 66.7 to 0% at 35 to 40 min, 0% at 40 to 45 min, 0 to 100% at 45 to 46 min, and 100% at 46 to 50 min.

3.6. Statistical Analysis

Each experimental result was obtained using three independent samples and is expressed as mean ± standard deviation (n = 3). For statistical analysis, a one-way ANOVA with Duncan's multiple range test was performed using IBM SPSS Statistics 20 (IBM Corp., Armonk, NY, USA), with significant differences set at $p < 0.05$.

4. Conclusions

In this study, the impact of ultrasonic frequencies on the aqueous extraction of bioactive compounds from WFP and RFP was investigated. The results demonstrated that U-40 was beneficial for enhancing the extraction efficiency of polyphenolic compounds from WFP and RFP but had different impacts. For WFP, the TPC value at 10 min was higher when using U-40 than when using U-120 and S-120 with significant differences, and it remained significantly higher at 60 min when using U-40 than when using U-120, indicating that U-40 and U-120 caused different levels of polyphenol degradation with a longer extraction time. For RFP, although the TPC values at 15 min were slightly higher when using U-40 than when using U-120 and S-120 with insignificant differences, U-40 was still beneficial for obtaining stable and higher polyphenol contents in RFP extracts. Additionally, the free CGA in WFP was more stable throughout the extraction duration when using U-120 than when using U-40, revealing that these methods have different impacts on the extraction of free CGA. When using S-120, severe degradation of the free CGA in WFP occurred after 15 min. Moreover, under the same extraction mode and conditions, the free quercetin contents in WFP were higher than those in RFP. The free quercetin content in RFP was significantly higher when using U-40 than when using S-120, showing that ultrasound effectively enhanced the extraction of hydrophobic free quercetin. The results of this study can be used as a reference when selecting the strategy for the green extraction of target bioactive compounds from whole pitaya fruit, promoting sustainability while simultaneously reducing waste generation, and green extraction could serve as potential technology to promote the value of pitaya for use in nutraceutical foods.

Author Contributions: Conceptualization, C.-Y.Y.; methodology, C.-Y.Y.; investigation, W.-T.L.; data curation, W.-T.L.; formal analysis, W.-T.L. and C.-Y.Y.; writing—original draft preparation, C.-Y.Y.; writing—review and editing, C.-Y.Y.; project administration, C.-Y.Y.; funding acquisition, C.-Y.Y. All authors have read and agreed to the published version of the manuscript.

Funding: This research was funded by the National Science and Technology Council, Taiwan, project No. NSTC 112-2221-E-030-001-MY3 awarded to Chun-Yao Yang.

Institutional Review Board Statement: Not applicable.

Informed Consent Statement: Not applicable.

Data Availability Statement: Data are contained within the article.

Acknowledgments: The authors thank the National Science and Technology Council, Taiwan (R.O.C.), for providing the financial support for this research (project No. NSTC 112-2221-E-030-001-MY3). The authors also thank Wen-Ling Hsiao for helping with the analysis of the moisture content.

Conflicts of Interest: The authors declare no conflicts of interest.

References

1. Jiang, H.; Zhang, W.; Li, X.; Shu, C.; Jiang, W.; Cao, J. Nutrition, phytochemical profile, bioactivities and applications in food industry of pitaya (*Hylocereus* spp.) peels: A comprehensive review. *Trends Food Sci. Technol.* **2021**, *116*, 199–217. [CrossRef]
2. Huang, Y.; Brennan, M.A.; Kasapis, S.; Richardson, S.J.; Brennan, C.S. Maturation process, nutritional profile, bioactivities and utilisation in food products of red pitaya fruits: A review. *Foods* **2021**, *10*, 2862. [CrossRef]
3. Arivalagan, M.; Karunakaran, G.; Roy, T.K.; Dinsha, M.; Sindhu, B.C.; Shilpashree, V.M.; Satisha, G.C.; Shivashankara, K.S. Biochemical and nutritional characterization of dragon fruit (*Hylocereus* species). *Food Chem.* **2021**, *353*, 129426. [CrossRef]
4. Joshi, M.; Prabhakar, B. Phytoconstituents and pharmaco-therapeutic benefits of pitaya: A wonder fruit. *J. Food Biochem.* **2020**, *44*, e13260. [CrossRef]
5. Tang, W.; Li, W.; Yang, Y.; Lin, X.; Wang, L.; Li, C.; Yang, R. Phenolic compounds profile and antioxidant capacity of pitahaya fruit peel from two red-skinned species (*Hylocereus polyrhizus* and *Hylocereus undatus*). *Foods* **2021**, *10*, 1183. [CrossRef] [PubMed]
6. Bishoyi, A.K.; Saeed, F.; Shehzadi, U.; Shankar, A.; Balaji, J.; Kaur, J.; Afzaal, M.; Imran, A.; Rasheed, M.; Hussain, B.; et al. Nutritional composition, phytochemical profile, and health benefits of *Hylocereus Undatus* (pitaya): A comprehensive review. *eFood* **2024**, *5*, e70017. [CrossRef]
7. Angonese, M.; Motta, G.E.; Silva de Farias, N.; Molognoni, L.; Daguer, H.; Brugnerotto, P.; de Oliveira Costa, A.C.; Olivera Müller, C.M. Organic dragon fruits (*Hylocereus undatus* and *Hylocereus polyrhizus*) grown at the same edaphoclimatic conditions: Comparison of phenolic and organic acids profiles and antioxidant activities. *LWT-Food Sci. Technol.* **2021**, *149*, 111924. [CrossRef]
8. Coelho, V.S.; de Moura, D.G.; Aguiar, L.L.; Ribeiro, L.V.; Silva, V.D.M.; da Veiga Correia, V.T.; Melo, A.C.; Silva, M.R.; de Paula, A.C.C.F.F.; de Araújo, R.L.B.; et al. The Profile of phenolic compounds identified in pitaya fruits, health effects, and food applications: An integrative review. *Plants* **2024**, *13*, 3020. [CrossRef]
9. Attar¸ S.H.; Günde, M.A.; Urün, I.; Kafkas, S.; Kafkas, N.E.; Ercisli, S.; Ge, C.; Mlcek, J.; Adamkova, A. Nutritional analysis of red-purple and white-fleshed pitaya (*Hylocereus*) Species. *Molecules* **2022**, *27*, 808. [CrossRef]
10. Santana-Gálvez, J.; Cisneros-Zevallos, L.; Jacobo-Velázquez, D.A. Chlorogenic acid: Recent advances on its dual role as a food additive and a nutraceutical against metabolic syndrome. *Molecules* **2017**, *22*, 358. [CrossRef]
11. Rasheed, H.; Ahmad, D.; Bao, J. Genetic diversity and health properties of polyphenols in potato. *Antioxidants* **2022**, *11*, 603. [CrossRef] [PubMed]
12. Aghababaei, F.; Hadidi, M. Recent advances in potential health benefits of quercetin. *Pharmaceuticals* **2023**, *16*, 1020. [CrossRef]
13. Deepika; Maurya, P.K. Health benefits of quercetin in age-related diseases. *Molecules* **2022**, *27*, 2498. [CrossRef]

14. Carmen, F.; Frances, C.; Barthe, L. Trends on valorization of pitaya fruit biomass through value-added and green extraction technology – A critical review of advancements and processes. *Trends Food Sci. Technol.* **2023**, *138*, 339–354. [CrossRef]
15. El-Hefny, N.E. Chemical kinetics and reaction mechanisms in solvent extraction: New trends and applications. *J. Phys. Sci.* **2017**, *28*, 129–156. [CrossRef]
16. Chen, Y.; Su, J.-Y.; Yang, C.-Y. Ultrasound-assisted aqueous extraction of chlorogenic acid and cynarin with the impact of inulin from burdock (*Arctium lappa* L.) roots. *Antioxidants* **2022**, *11*, 1219. [CrossRef] [PubMed]
17. Passos, C.P.; Costa, R.M.; Ferreira, S.S.; Lopes, G.R.; Cruz, M.T.; Coimbra, M.A. Role of coffee caffeine and chlorogenic acids adsorption to polysaccharides with impact on brew immunomodulation effects. *Foods* **2021**, *10*, 378. [CrossRef] [PubMed]
18. Zhong, X.; Zhang, S.; Wang, H.; Yang, J.; Li, L.; Zhu, J.; Liu, Y. Ultrasound-alkaline combined extraction improves the release of bound polyphenols from pitahaya (*Hylocereus undatus* 'Foo-Lon') peel: Composition, antioxidant activities and enzyme inhibitory activity. *Ultrason. Sonochem.* **2022**, *90*, 106213. [CrossRef]
19. Tarte, I.; Singh, A.; Dar, A.H.; Sharma, A.; Altaf, A.; Sharma, P. Unfolding the potential of dragon fruit (*Hylocereus* spp.) for value addition: A review. *eFood* **2023**, *4*, e76. [CrossRef]
20. Nutrition Information Database, Pitaya: D0700101 (White-flesh), D0700201 (Red-flesh), 2024UPDATE2, Food and Drug Administration, Taiwan (R.O.C.). Available online: https://consumer.fda.gov.tw/Food/TFND.aspx?nodeID=178 (accessed on 13 July 2025).
21. Wu, L.C.; Hsu, H.W.; Chen, Y.C.; Chiu, C.C.; Lin, Y.I.; Ho, J.A. Antioxidant and antiproliferative activities of red pitaya. *Food Chem.* **2006**, *95*, 319–327. [CrossRef]
22. Smirnov, I.; Mikhailova, N. An analysis of acoustic cavitation thresholds of water based on the incubation time criterion approach. *Fluids* **2021**, *6*, 134. [CrossRef]
23. Nguyen, T.T.; Asakura, Y.; Koda, S.; Yasuda, K. Dependence of cavitation, chemical effect, and mechanical effect thresholds on ultrasonic frequency. *Ultrason. Sonochem.* **2017**, *39*, 301–306. [CrossRef] [PubMed]
24. Bu, X.; Alheshibri, M. The effect of ultrasound on bulk and surface nanobubbles: A review of the current status. *Ultrason. Sonochem.* **2021**, *76*, 105629. [CrossRef] [PubMed]
25. Klöckner, W.; Büchs, J. Advances in shaking technologies. *Trends Biotechnol.* **2012**, *30*, 307–314. [CrossRef]
26. Wang, P.; Cheng, C.; Ma, Y.; Jia, M. Degradation behavior of polyphenols in model aqueous extraction system based on mechanical and sonochemical effects induced by ultrasound. *Sep. Purif. Technol.* **2020**, *247*, 116967. [CrossRef]
27. Kumar, K.; Srivastav, S.; Sharanagat, V.S. Ultrasound assisted extraction (UAE) of bioactive compounds from fruit and vegetable processing by-products: A review. *Ultrason. Sonochem.* **2021**, *70*, 105325. [CrossRef] [PubMed]
28. Su, D.; Zhang, R.; Hou, F.; Zhang, M.; Guo, J.; Huang, F.; Deng, Y.; Wei, Z. Comparison of the free and bound phenolic profiles and cellular antioxidant activities of litchi pulp extracts from different solvents. *BMC Complement. Altern. Med.* **2014**, *14*, 9. [CrossRef] [PubMed]
29. Yeh, Y.M.; Tsai, T.Y.; Yang, C.Y. Encapsulation and release kinetics of polyphenols and *p*-coumaric acid extracted from *Phyllostachys makinoi* by ultrasonic-pretreatment autoclaving. *LWT-Food Sci. Technol.* **2022**, *167*, 113838. [CrossRef]
30. Razmara, R.S.; Daneshfar, A.; Sahraei, R. Solubility of quercetin in water + methanol and water + ethanol from (292.8 to 333.8) K. *J. Chem. Eng. Data* **2010**, *55*, 3934–3936. [CrossRef]
31. Hsu, Y.-C.; Yang, C.-Y. Assessment of ethanolic extraction of chlorogenic acid, cynarin, and polyphenols from burdock (*Arctium lappa* L.) roots under ultrasound. *Molecules* **2024**, *29*, 5115. [CrossRef]
32. Kujala, T.S.; Loponen, J.M.; Klika, K.D.; Kalevi, P. Phenolics and betacyanins in red beetroot (*Beta vulgaris*) root: Distribution and effect of cold storage on the content of total phenolics and three individual compounds. *J. Agric. Food Chem.* **2000**, *48*, 5338–5342. [CrossRef] [PubMed]
33. Madane, P.; Das, A.K.; Nanda, P.; Bandyopadhyay, S.; Jagtap, P.; Shewalkar, A.; Maity, B. Dragon fruit (*Hylocereus undatus*) peel as antioxidant dietary fibre on quality and lipid oxidation of chicken nuggets. *J. Food Sci. Technol.* **2020**, *57*, 1449–1461. [CrossRef] [PubMed]
34. Chen, Y.H.; Yang, C.Y. Ultrasound-assisted extraction of bioactive compounds and antioxidant capacity for the valorization of *Elaeocarpus serratus* L. leaves. *Processes* **2020**, *8*, 1218. [CrossRef]

molecules

Article

Optimization and Impact of Ultrasound-Assisted Extraction on Pomegranate Seed Oil Quality: A Comparative Study of Bioactive Potential and Oxidation Parameters

Marta Siol, Iga Piasecka, Diana Mańko-Jurkowska, Agata Górska and Joanna Bryś *

Department of Chemistry, Institute of Food Sciences, Warsaw University of Life Sciences, Nowoursynowska St. 159c, 02-787 Warsaw, Poland; marta_siol@sggw.edu.pl (M.S.); iga_piasecka@sggw.edu.pl (I.P.); diana_manko-jurkowska@sggw.edu.pl (D.M.-J.); agata_gorska@sggw.edu.pl (A.G.)
* Correspondence: joanna_brys@sggw.edu.pl

Abstract: Pomegranate seed oil (PSO), a by-product of juice production, is rich in bioactive compounds, especially punicic acid, and has significant potential for health and industrial applications. The present study aimed to optimize an ultrasound-assisted extraction (UAE) of PSO and compare its effectiveness with conventional methods such as cold pressing and Soxhlet extraction. A Box–Behnken design was used to determine the optimal UAE parameters (amplitude 46%, 12 min, L/S ratio 19 mL/g), yielding 12.67% oil with the highest oxidative stability (τ_{max} = 5.44 min). Compared to Soxhlet and cold-pressed methods, UAE gave the highest yield, but slightly lower levels of total polyphenols and antioxidant activity. Cold-pressed oil retained the most bioactive compounds, but showed reduced oxidative stability and higher susceptibility to degradation. Soxhlet extraction provided moderate antioxidant capacity and the highest punicic acid content, but exceeded the recommended limits for acid value. Overall, the UAE offers an effective balance between yield, quality, and sustainability, with minimal thermal degradation and reduced solvent consumption. The results confirm that UAE is a promising alternative for high-quality PSO extraction, although cold pressing remains superior in preserving sensitive bioactive components. Ultimately, this study underscores that the extraction method plays a decisive role in determining the functional quality and oxidative stability of PSO, with UAE standing out as the most efficient and environmentally favorable approach.

Keywords: pomegranate seed oil; ultrasound-assisted extraction; antioxidant activity; antioxidant stability; oxidation kinetics parameters; fatty acid profile

1. Introduction

Fruit seeds are a significant by-product generated in large quantities during fruit processing. Accounting for approximately 20% of the fruit's total weight, pomegranate seeds are a significant by-product of juice production. However, due to limited research and valorization, they contribute to an estimated annual waste of 70,000–100,000 tons [1,2].

Despite being a rich source of bioactive compounds, fruit seeds are often discarded as waste, leading to both the loss of valuable raw materials and an increased environmental burden [3–5]. However, scientific studies indicate that seeds, including those of pomegranates, contain essential nutrients and bioactive components that, when appropriately processed, can have diverse applications [6,7]. Among the most valuable products derived from pomegranate seeds is pomegranate seed oil (PSO). The extraction of PSO not only facilitates the utilization of this valuable resource but also aligns with the principles of a circular economy and the sustainable development strategy advocated by the European Union policy.

The oil present in pomegranate seeds is gaining increasing interest due to its unique fatty acid (FA) composition. PSO is distinguished by a high content of unsaturated fatty acids and conjugated fatty acids, among which the most significant is the isomer of α-linolenic acid—punicic acid (9-*cis*, 11-*trans*, 13-*cis* octadecatrienoic acid) [8,9]. Punicic acid (PA) is known for its diverse health benefits, including anticancer and anti-inflammatory properties. Studies indicate that PA exhibits selective cytotoxicity against cancer cells while sparing normal cells, highlighting its potential as a therapeutic agent [10]. In addition, PA has been shown to improve metabolic disorders by regulating the gut microbiota and reducing lipid accumulation in the liver [11]. Moreover, it has shown antiatherosclerotic and anti-diabetic effects and may be useful in combating insulin resistance and even in treating type 2 diabetes [12–14]. Furthermore, PSO is rich in bioactive compounds, mainly polyphenols and tocopherols, which exhibit antioxidant, anti-inflammatory, and antimicrobial properties. Polyphenols, mainly ellagic acid and anthocyanins, are antioxidants that help neutralize free radicals, reducing oxidative stress and inflammation [15]. The high levels of polyphenols and tocopherols in PSO provide potent antioxidant activity, crucial for preventing chronic diseases [15]. Additionally, the bioactive compounds in PSO have been shown to reduce inflammation and exhibit antimicrobial properties. Notably, PSO has effectively decreased mycotoxin secretion, highlighting its potential as a natural preservative in food systems [16]. The choice of oil extraction methods is critical to optimizing yield and preserving health-promoting components. Efficiently extracting PSO while preserving its bioactive components presents several challenges, as traditional methods such as solvent extraction and hot pressing can weaken the bioactive components of PSO due to thermal degradation [6].

Cold pressing is a method widely used in industry because of its ability to preserve bioactive compounds. However, compared to extraction with solvents, this method is much less efficient and more time consuming [17]. In addition, cold-pressed oils have a shorter shelf life. This is because they retain more natural compounds and do not contain chemical preservatives, and are more susceptible to oxidation and rancidity, requiring careful storage conditions [18]. Despite these limitations, cold pressing remains popular because of its environmental friendliness and the high quality of the oil that is produced, free of chemical residues. Soxhlet extraction is highly effective for extracting bioactive compounds from solid materials. Compared to cold pressing, in the Soxhlet method, more oil can be extracted due to the repeated recycling of solvent. However, prolonged exposure to heat can degrade thermolabile (heat-sensitive) compounds, affecting the quality of the extract [19]. Soxhlet extraction is identified as optimal for laboratory conditions, yielding high-quality oil, but is not widely used in industry, mainly due to oil contamination by residual solvent used in extraction [20].

Ultrasound-assisted extraction (UAE) is a technique that enhances the extraction of bioactive compounds from plant materials by utilizing ultrasonic waves. This method is known for its efficiency in breaking down cell walls and improving mass transfer, leading to higher yields of extracted substances. UAE works through the phenomenon of cavitation, in which ultrasonic waves create microbubbles in the solvent. These bubbles grow and implode, generating intense local pressure and temperature changes that break down cell walls, facilitating the release of intracellular compounds [21]. Implosion of cavitation bubbles increases solvent penetration into plant tissues, further aiding the extraction process [21]. Ultrasonic waves induce micro-acoustic fluxes and pressure differences around the particles, significantly increasing the mass transfer rate. UAE is distinguished by reduced energy consumption and minimal solvent consumption compared to conventional extraction methods. This makes it a more sustainable option for extracting bioactive compounds [22]. To maximize efficiency and yield, the process can be optimized by adjusting parameters such as ultrasound power, temperature, and solvent type [23].

The purpose of this study is to optimize the extraction process of PSO using UAE and to evaluate the effect of this method on FA composition, oxidative stability, quality parameters, and bioactive compound content compared to conventional oil obtaining methods such as cold pressing and Soxhlet extraction. The study aims to determine whether UAE is an effective alternative to conventional techniques in terms of the quality and health-promoting properties of the obtained oil.

2. Results and Discussion

2.1. Experimental Design—Ultrasound-Assisted Extraction

The UAE variables' influence on yield and τ_{max} responses was analyzed. The suitable models and equations were fitted based on ANOVA results (Table 1). For the yield, a linear model was chosen as significant ($p \leq 0.05$), and for τ_{max}, a quadratic model ($p \leq 0.05$) was preferred. Graphical expressions of the surface response are shown in Figure 1.

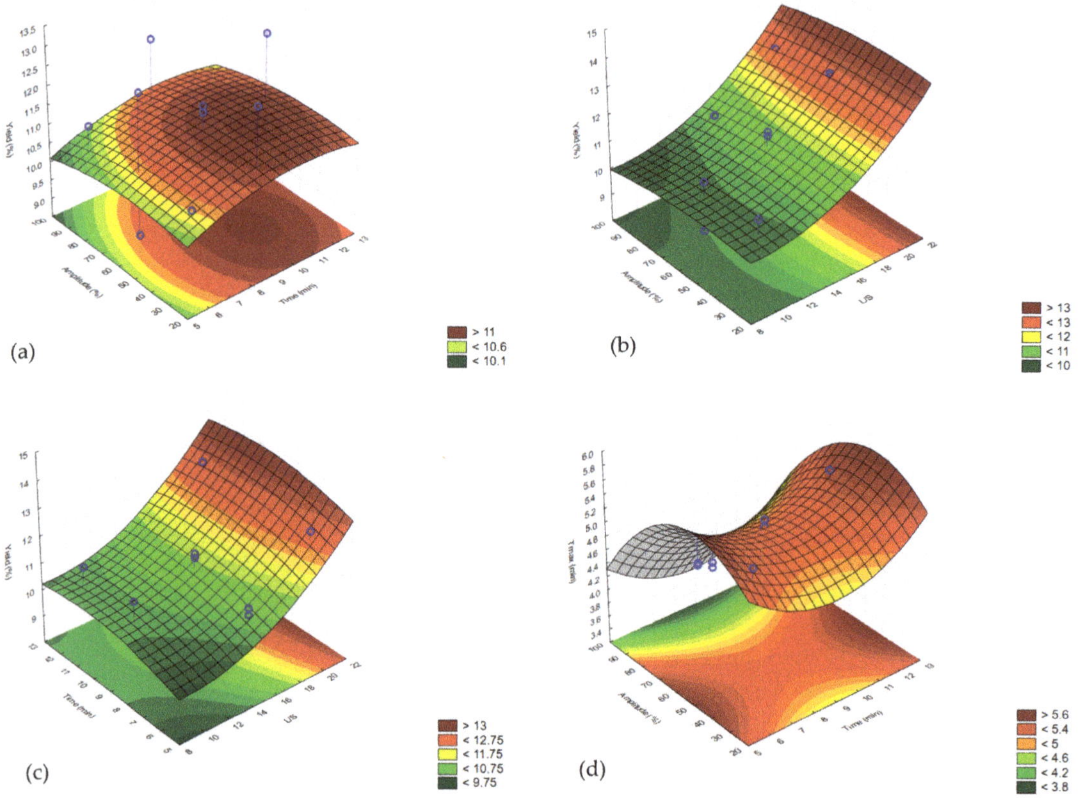

Figure 1. *Cont.*

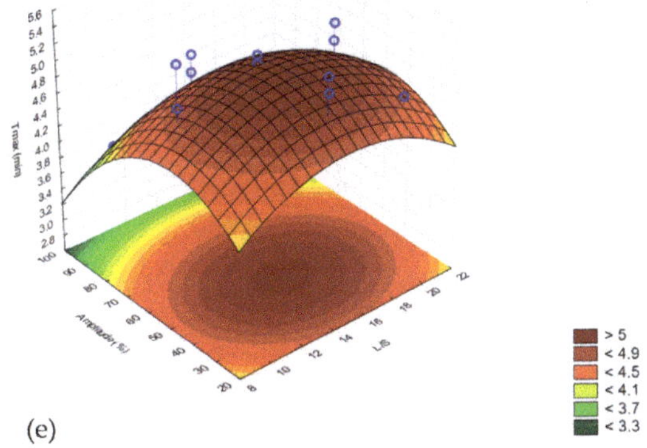

(e)

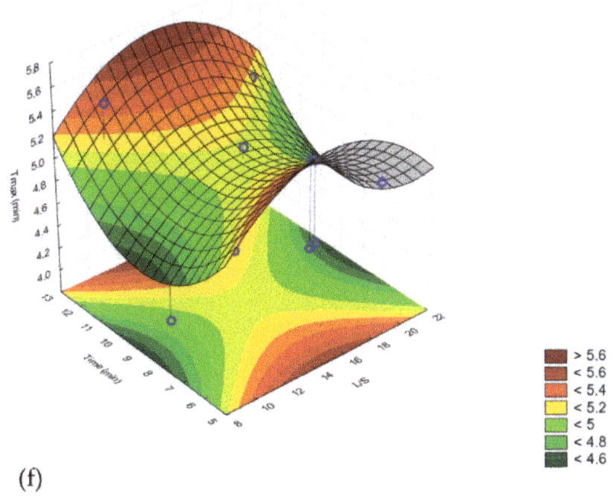

(f)

Figure 1. Effect of (**a**) extraction time and ultrasonic amplitude on oil yield at liquid to solid (L/S) ratio of 15; (**b**) ultrasonic amplitude and L/S ratio on oil yield at 9 min; (**c**) time and L/S ratio on oil yield at 60% ultrasonic amplitude; (**d**) extraction time and ultrasonic amplitude on maximum oxidation time (τ_{max}) at L/S ratio of 15; (**e**) ultrasonic amplitude and L/S ratio on τ_{max} at 9 min; (**f**) time and L/S ratio on τ_{max} at 60% ultrasonic amplitude.

Table 1. Results of ANOVA for Box–Behnken design model and equation fitting for ultrasound-assisted pomegranate seed oil extraction.

Response	Model Fitted	R^2	CV (%)	Model *p*-Value	Lack of Fit *p*-Value
Yield	Linear	0.7544	5.45	0.0011	0.7319
τ_{max}	Quadratic	0.9201	4.45	0.0274	0.1146

where τ_{max}—maximum oxidation time, R^2—determination coefficient, CV—coefficient of variation.

Generally, for the extraction yield, the only significant ($p \leq 0.05$) variable was the L/S ratio. The observed relationship indicated that an increase in the L/S ratio corresponded to an enhancement in extraction yield, reaching the highest value of 12.85% in run number 8 (See section: Materials and Methods). The exact dependence is explained in Equation (1). The results obtained in previous findings are inconsistent in terms of factors influencing extraction yield. Tian et al. [24] also reported that the higher the L/S ratio, the higher the PSO yield could be obtained in the UAE process. According to the results obtained by Barizão et al. [25] in the UAE, L/S ratio, time, and temperature had a significant effect on PSO yield, however, the dependency of yield increase with increasing L/S ratio was confirmed only to some extent. When the L/S ratio reached a maximum of 28 mL/g, the yield was decreased compared to the results obtained for a 24 mL/g L/S ratio.

$$Yield = 35.77 + 0.33 \cdot Time - 0.02 \cdot Amplitude + 1.24 \cdot L/S \quad (1)$$

In terms of maximum oxidation time (τ_{max}), which represents resistance to oxidation of the oil sample, the square of time and amplitude, and also amplitude itself were identified as significant ($p \leq 0.05$) variables. The results of τ_{max} were in a range of 3.95–5.47 min. Rather longer sonication (12 min) accompanied by mild amplitude values (60%) resulted in longer τ_{max}. Based on the results, the adequate model described by Equation (2) was fitted. Maximum oxidation time has not been assessed in the optimization studies for UAE of pomegranate seed oil yet. However, PSO is recognized as prone to oxidation reactions due to a high abundance of unsaturated fatty acids [26], and any improvement noted with

UAE application is remarkable. As the UAE enables obtaining the products in a short treatment time, its application may prevent oil degradation when compared to traditional solid–liquid extraction. It is highly important, especially regarding oil with a significant amount of polyunsaturated fatty acids (PUFAs). Also, the lower temperature of the UAE prevents bioactive compounds from deteriorating. However, if the sonication is too intense or too long, it may result in peroxide formation and promote the oxidation of oils. That is why there is a need to optimize the UAE procedure to find the most effective extraction conditions that do not cause extensive peroxide formation [27].

$$\begin{aligned}\tau_{max} = 3.39 - 0.59 \cdot Time - 0.06 \cdot Amplitude + 0.40 \cdot L/S - 0.00008 \cdot Time \\ \cdot Amplitude - 0.005 \cdot Time \cdot L/S - 0.003 \cdot Amplitude \cdot L/S \\ + 0.04 \cdot Time^2 - 0.0005 \cdot Amplitude^2 + 0.01 \cdot L/S^2\end{aligned} \quad (2)$$

The results were analyzed in ANOVA to obtain the most effective conditions. The most optimal process parameters were selected to maximize both extraction yield and maximum oxidation time (τ_{max}) to the greatest extent possible. The extraction conditions were set within the range proposed in the experiment, specifically an amplitude of 30–90%, 6–12 min durations, and an L/S ratio of 10–20. The software proposed 38 solutions, and one with the highest desirability was chosen. The exact values were 45.82% amplitude, 12 min time, and 18.88 L/S ratio. To adjust the proposed variables, the real applied parameters were 46% amplitude, 12 min time, and 19 L/S ratio. The predicted responses of yield and τ_{max} were 12.43% and 5.30, respectively. The mean yield result was 12.67 ± 0.48%, which differs from the predicted value by only 1.9%. For the τ_{max}, the actual mean value of 5.44 ± 0.02 min was recorded, which deviates from the predicted one by 2.6%. The results indicate that the model and equation predicted the results well.

2.2. Oil Yield

Each obtaining method has unique advantages and efficiencies, affecting the quality and composition of the final oil. The yield of the process of obtaining PSO varied significantly depending on the applied extraction method (Table 2).

Table 2. Pomegranate seed oil extraction yield dependence on the applied method.

Oil	Yield [%]
CP_PSO	9.61 [a] ± 0.22
SE_PSO	11.49 [b] ± 0.19
UAE_PSO	12.67 [c] ± 0.48

CP_PSO—cold-pressed pomegranate seed oil; SE_PSO—pomegranate seed oil extracted by the Soxhlet method; UAE_PSO—pomegranate seed oil extracted by ultrasound-assisted extraction. The different lower-case letters indicate significantly different values ($p \leq 0.05$). Data are presented as mean values followed by standard deviation (±SD).

The lowest yield was observed for cold-pressed oil (CP_PSO), with a recovery of 9.61 ± 0.22%. The Soxhlet extraction method (SE_PSO) exhibited a higher yield, reaching 11.49 ± 0.19%, indicating its greater efficiency compared to cold pressing. The highest yield was obtained using ultrasound-assisted extraction, with a value of 12.67 ± 0.48%. The superior efficiency of the Soxhlet and UAE methods may be attributed to the enhanced lipid extraction in the presence of solvents and/or the application of more intensive process conditions, such as elevated temperature and ultrasound treatment. In contrast, the lower yield of cold pressing suggests that although this method is more beneficial for preserving bioactive compounds, it is less efficient in terms of oil recovery.

The obtained extraction yields are consistent with previous literature reports, which highlight the significant impact of extraction parameters on the amount of oil recovered from pomegranate seeds. In the study by Dib et al. [28], the Soxhlet extraction yield of PSO using *n*-hexane as a solvent ranged from 15.62% to 16.25%, which is slightly higher than the yield obtained in this study. Similarly, in the study by Tian et al. [24], the Soxhlet extraction yield was reported as 20.50%. Regarding cold pressing, the research by Kaseke et al. [29] reported a yield of 3.20%, which is approximately three times lower than the yield observed in this study. Furthermore, previous studies on the optimization of pomegranate seed oil extraction using UAE reported higher yields than those obtained in the present study, reaching 25.11% [24] or 27.99% [25].

However, the content of oil in the pomegranate seeds may be influenced by the origin of the seeds, the cultivar, and moisture content. In the case of cold pressing, the efficiency of oil extraction may also depend on the oil press equipment used in the process [29].

2.3. Fatty Acid Profile of Oils

In vegetable oils, the predominant group of lipids are triacylglycerols (TAGs). Both the composition and distribution of FA in TAG affect the nutritional properties and stability of the oil. Table 3 shows the FA profile of PSO obtained by different methods. Regardless of the method of oil extraction, the same FAs in similar contents were identified. In PSO, saturated fatty acids (SFAs) accounted for approximately 6%, while monounsaturated fatty acids (MUFAs) contributed around 5–6%. The FA composition of PSO was dominated by PUFAs, especially punicic acid (C18:3 n-5), and its isomers, which together accounted for approx. 85%. The slightly lower content of PA in the UAE-treated samples (78.68%) compared to the Soxhlet (79.15%) ones and significantly lower than in cold-pressed (80.80%) samples, despite a higher extraction yield, may suggest the occurrence of subtle structural modifications, such as partial degradation or isomerization, as a result of ultrasonic treatment. Although the observed difference is relatively minor and may fall within the range of analytical or biological variability, it is consistent with previous studies reporting that high-energy ultrasound can affect the integrity of conjugated systems under certain conditions. In particular, thermolabile FAs like PA may be susceptible to changes induced by localized hotspots, shear forces, and the formation of free radicals during sonication, potentially leading to the generation of positional or geometrical isomers, or even oxidative degradation products [30,31].

Table 3. Fatty acid profile [%] of pomegranate seed oils.

	CP_PSO	SE_PSO	UAE_PSO
C16:0	$2.45^{b} \pm 0.06$	$2.27^{a} \pm 0.02$	$2.42^{b} \pm 0.02$
C18:0	$2.05^{b} \pm 0.03$	$1.90^{a} \pm 0.02$	$1.99^{ab} \pm 0.04$
C18:1 n-9c	$4.96^{c} \pm 0.08$	$4.54^{a} \pm 0.04$	$4.75^{b} \pm 0.06$
C18:2 n-6c	$4.72^{b} \pm 0.08$	$4.42^{a} \pm 0.04$	$4.51^{a} \pm 0.05$
C20:0	$0.54^{a} \pm 0.02$	$0.51^{a} \pm 0.01$	$0.49^{a} \pm 0.03$
C20:1 n-9c	$0.73^{a} \pm 0.01$	$0.75^{a} \pm 0.01$	$0.71^{a} \pm 0.01$
C18:3 (9c, 11t, 13c)	$80.80^{b} \pm 0.33$	$79.15^{a} \pm 0.08$	$78.68^{a} \pm 0.21$
isomers of C18:3 (9c, 11t, 13c)	$3.48^{a} \pm 0.07$	$6.23^{b} \pm 0.17$	$6.21^{b} \pm 0.17$
other	$0.28^{a} \pm 0.01$	$0.26^{a} \pm 0.01$	$0.27^{a} \pm 0.13$
Σ SFA	5.04^{c}	4.67^{a}	4.89^{b}
Σ MUFA	5.69^{c}	5.28^{a}	5.46^{b}
Σ PUFA	86.56^{b}	89.80^{c}	84.88^{a}

CP_PSO—cold-pressed pomegranate seed oil; SE_PSO—pomegranate seed oil extracted by the Soxhlet method; UAE_PSO—pomegranate seed oil extracted by ultrasound-assisted extraction SFA—saturated fatty acids; MUFA—monounsaturated fatty acids; PUFA—polyunsaturated fatty acids; The different lower-case letters show a statistically significant difference in the row at a $p \leq 0.05$. Data are presented as mean values followed by standard deviation (±SD).

The highest percentage of PUFAs was obtained for oil extracted using the Soxhlet method, and the lowest for oil obtained using ultrasound treatment.

The literature data indicate that PA is the characteristic FA of PSO, accounting for 55 to 81% of all FAs. The composition of FAs (including the content of PA) can vary depending on the pomegranate variety, the place it was grown, the time it was harvested, or the method the oil was obtained [32]. Rowayshed et al. [33] studied PSO extracted using petroleum ether in a Soxhlet apparatus, which included only 59.40% PA. Habibnia et al. [34] showed that the oil, extracted from five different Iranian pomegranate seed varieties using *n*-hexane as the solvent at low temperatures, contained between 78.25 and 82.40% PA. Their data were similar to those obtained for PSO in the present study. Dadashi et al. [35] also extracted oil using *n*-hexane at room temperature, and they determined lower PA contents (from 72.07% to 73.31%) in the PSO tested than Habibnia et al. [34]. Juhaimi et al. [36] extracted oil from six fruit varieties of pomegranate picked from the Mediterranean region in Turkey, including the Hicaz one, as in the present study. A unique PA was found in pomegranate seed as the dominant FA in the range from 71.17 to 77.62%, while the highest value was noted for the Hicaz variety. Eikani et al. [37] compared superheated hexane extraction (SHHE), Soxhlet extraction, and the cold-pressing method to extract PSO. The FA profile for SHHE was more similar to that obtained by the cold-pressing method. Moreover, the percentage of PA was significantly lower after SHHE or cold pressing (70.73 and 69.79%, respectively) than after Soxhlet extraction (81.69%). Such significant differences in the percentage content of PA, depending on the method of oil obtaining, were not noted in the present study.

2.4. *Quality Parameters of Oils*

The acid value (AV) determines the amount of free fatty acids in fat and is expressed in mg KOH per 1 g of fat. According to the standards established by the Codex Alimentarius [38], AV should not exceed 4 mg KOH/g of fat in cold-pressed oils. The peroxide value (PV) is a measure of the peroxide content, which is the primary product of lipid oxidation, and indicates the degree of oxidation or rancidity of fat. It is expressed in milliequivalents of active oxygen per 1 kg of fat. The PV for cold-pressed oils is specified in the EN ISO 3960:2017-03 standard [39], which states that it should not exceed 10 meq O_2/kg, while the Codex Alimentarius [38] sets the limit at 15 meq O_2/kg. Table 4 shows the quality parameters of the extracted PSO.

Table 4. Quality parameters of pomegranate seed oils.

Oil	AV [mg KOH/g]	PV [meq O_2/kg]	p-AnV	TOTOX (2PV + p-AnV)
CP_PSO	2.23 ^{a} ± 0.10	4.68 ^{b} ± 0.15	15.78 ^{b} ± 0.21	25.14 ^{c} ± 0.21
SE_PSO	4.75 ^{c} ± 0.16	3.20 ^{a} ± 0.24	14.76 ^{a} ± 0.14	21.17 ^{a} ± 0.26
UAE_PSO	3.08 ^{b} ± 0.22	4.52 ^{b} ± 0.21	14.91 ^{a} ± 0.28	23.95 ^{b} ± 0.34

CP_PSO—cold-pressed pomegranate seed oil; SE_PSO—pomegranate seed oil extracted by the Soxhlet method; UAE_PSO—pomegranate seed oil extracted by ultrasound-assisted extraction; AV—acid value; PV—peroxide value; p-AnV—anisidine value. The different lower-case letters indicate significantly different values ($p \leq 0.05$). Data are presented as mean values followed by standard deviation (±SD).

The results indicate that cold-pressed PSO exhibited the lowest AV among the tested samples (2.23 mg KOH/g). The highest AV was observed for PSO obtained by the Soxhlet method (4.75 mg KOH/g), which exceeded the recommended standard. Contrarily, the lowest PV was recorded for oil extracted using the Soxhlet method (3.20 meq O_2/kg), while all tested oils complied with the 10 meq O_2/kg fat limit established by EN ISO 3960:2017-03 [39]. The PV of cold-pressed PSO (4.68 meq O_2/kg) was the highest among the tested oils.

The literature data show a wide range of AV and PV obtained for PSO, as they depend on many factors, including the origin and quality of the raw material, its cultivation conditions, methods, and parameters of oil production or storage conditions. Khoddami and Roberts [40] reported that the PV of cold-pressed PSO ranged from 4.67 to 5.96 meq O_2/kg fat, and the AV value of CP_PSO (4.68 meq O_2/kg fat) falls within this range. Meanwhile, Juhaimi et al. [36] obtained higher PV values (5.44 to 8.47 meq O_2/kg fat) for PSO and Costa et al. [41]—lower ones (0.91–2.69 meq O_2/kg fat). For the commercial PSO studied by Melo et al. [8], the AV (0.63 mg KOH/g fat) was significantly lower than for PSO presented in this study, but the PV (3.34 meq O_2/kg) was close to the PV of SE_PSO (3.20 meq O_2/kg). In turn, Dadashi et al. [35] determined AV values ranging from 3.78 to 8.36 mg KOH/g fat for Soxhlet-extracted PSO, similar to the AV of SE_PSO (4.75 mg KOH/g fat) in the present study but at the same time—much lower PV (0.39–0.50 meq O_2/kg fat).

During storage, peroxides can undergo further transformations, leading to the formation of secondary oxidation products such as aldehydes or ketones, which affect undesirable changes in the taste, odor, and nutritional value of the oil. The anisidine value (p-AnV) determines the content of secondary oxidation products, primarily aldehydes. It is considered a more sensitive indicator of advanced fat oxidation than PV. Considering the level of secondary oxidation products, comparable results were obtained for extracted oils and higher for cold-pressed PSO. Also, the TOTOX index, which provides information on both early and advanced stages of fat oxidation, was the highest for CP_PSO (25.14). These findings suggest that cold-pressed oil may be more prone to oxidation, likely due to enhanced exposure to oxygen during processing.

In the study by Ghorbanzadeh and Rezaei [42], cold-pressed PSO exhibited a significantly lower p-AnV (5.11) and TOTOX index (11.61) compared to PSO obtained by the same method in the present study. Conversely, the p-AnV (17.17) and TOTOX (26.7) values for oil extracted using the Soxhlet method were slightly higher than those obtained in this study [42]. The results obtained for PSO extracted by the Soxhlet method in the present study demonstrated higher p-AnV and TOTOX compared to the findings reported by Alaşalvar et al. [43], in which the p-AnV was 8.54, and the TOTOX index reached the value of 17.52. Although the UAE is increasingly employed for the extraction of plant oils due to its high efficiency, shortened processing time, and environmental friendliness, the scientific literature still lacks comprehensive data concerning such quality indicators as p-AnV and TOTOX index for PSO extracted using this method.

Higher values of PV, p-AnV, and TOTOX in cold-pressed PSO compared to those obtained by Soxhlet extraction or UAE can be due to several factors: (1) cold-pressed oil is obtained through mechanical pressing at low temperature without any refining step, therefore, it can contain not only naturally occurring antioxidants but also more oxidation-prone compounds, which can contribute to higher PV and p-AnV; (2) cold pressing involves prolonged contact with air, leading to oxidation of PUFA; (3) Soxhlet extraction and UAE use organic solvents or cavitation effects, respectively; therefore, they could provide better lipid solubilization and reducing oxidation by limiting contact with air during the process; (4) UAE operates under controlled conditions that minimizes oxidation. The observed differences in quality parameters suggest that the method of fat obtaining may influence the oxidative stability of PSO.

2.5. Oxidation of Oils

2.5.1. Oxidative Stability

Oxidative stability, or oil's resistance to oxidation processes, is a key parameter of edible oils that determines their quality, shelf life, nutritional value, and suitability for consumption and processing [44]. The pressure differential scanning calorimetry (PDSC) method provides a rapid and reproducible assessment of lipid oxidation in the selected oil under accelerated conditions, making it a highly effective tool for oxidative stability analysis. This technique enables the real-time recording of oxidation changes in a PDSC oxidation curve (heat flow changes as a function of time), allowing for precise characterization of the oil's oxidative behavior, e.g., for determining oxidation-related parameters like oxidation induction time (OIT). OIT is the time from the initial exposure to oxygen to the onset of exothermic decomposition at the isothermal conditions of the test. It is known that the longer the OIT is, the more stable the oil is [45].

Based on the obtained PDSC oxidation curves, both OIT and the oxidation time corresponding to the maximum heat flow values (to the maximum intensity of oxidative changes occurring in the sample; τ_{max}) were determined and shown in Table 5. OIT and τ_{max} decreased with increasing oxidation temperature, regardless of the method in which the oil was obtained. It is a well-known tendency, as, according to van 't Hoff's empirical rule, when the temperature increases by approximately 10 K, the reaction rate increases approx. 2–4 times [46]. The present study indicated that increasing the temperature by 10 °C between 90 and 120 °C reduced OIT by about 2.6 to 3 times. The temperature coefficient, which determines how many times the reaction rate increases when the temperature is raised, was the highest for cold-pressed PSO. Oxidative changes also began much faster in cold-pressed PSO (e.g., ~60 min. at 90 °C) than in oil extracted by the Soxhlet method or by ultrasound (~105–107 min. at 90 °C). As was mentioned earlier, cold pressing is a mechanical process that does not involve using solvents or high temperatures; therefore, the obtained oil can contain more impurities, including enzymes, trace metals, phospholipids, free fatty acids, and moisture, than oil after Soxhlet extraction. Such impurities can act as oxidation catalysts or pro-oxidants and accelerate oxidation. However, in the present study, it was shown that the AV value of CP_PSO was lower than that of SE_PSO; therefore, the amount of free fatty acids was lower in cold-pressed PSO. In turn, SE_PSO was characterized by lower PV, p-AnV, and TOTOX than other tested oils. It can be stated that the oil has undergone hydrolytic breakdown rather than oxidative damage. Cold-pressed oil preserves natural enzymes from the seeds, such as lipoxygenase and peroxidase, which can accelerate lipid oxidation; meanwhile, the Soxhlet method may inactivate these enzymes, leading to slower oxidation rates. Moreover, cold pressing preserves all naturally occurring compounds, but some of them may contribute to pro-oxidation reactions when exposed to oxygen, while Soxhlet extraction can selectively remove some oxidation-prone minor compounds, increasing overall stability. It is worth mentioning that cold-pressing exposes the oil to oxygen and friction-generated heat, which initiates oxidation at an early stage, but Soxhlet extraction occurs in a closed system with a solvent, minimizing direct exposure to oxygen and delaying the onset of oxidation.

A study by Yoshime et al. [47] showed that PSO was characterized by a shorter OIT than other edible oils. According to the authors, it was related to the high concentration of PA, which dominated the composition of FA. It is known that a significant number of multiple bonds increases the rate of oil oxidation; for example, linoleic acid oxidizes from 10 to 40 times faster than oleic acid, whereas α-linolenic acid oxidizes 2–4 times faster than linoleic acid [48]. PA with three double bonds is an isomer of α-linolenic acid; however, in PA, these double bonds are conjugated rather than methylene interrupted, which alters its reactivity. Ratusz et al. [49] analyzed the autooxidation process of model samples of refined vegetable oils, such as rapeseed and sunflower, using PDSC. The oxidative stability was determined at 120 °C according to the methodology used in the present paper. Their results showed that OIT for rapeseed oil was 56.73 min, while for sunflower oil, it reached 24.55 min. Since PSO belongs to the group of specialty oils, the stability of PSO was also compared with data available in the literature for such kind of oil, e.g., watermelon seed oil. The lowest τ_{max} value for unrefined cold-pressed watermelon seed oil (23.88 min) determined at 120 °C by Siol et al. [50] was more than 7× higher than for CP_PSO (3.37 min). In contrast, the authors obtained the highest τ_{max} value for watermelon seed oil extracted with hexane at room temperature (76.55 min). The difference between τ_{max} values for cold-pressed and extracted watermelon seed oil was twice as much as for PSO determined in this study at the same temperature. Most likely, depending on whether it is the FA composition or the antioxidant content that determines to a greater extent the oxidative stability of a given oil, its method of extraction may have a greater or lesser effect on the observed differences in oxidative stability.

Table 5. Oxidation induction time (OIT) and oxidation time corresponding to the maximum intensity of oxidative changes (τ_{max}) in pomegranate seed oils. Data were obtained at different temperatures.

Oil	90 °C		100 °C		110 °C		120 °C	
	OIT [min]	τ_{max} [min]	OIT [min]	τ_{max} [min]	OIT [min]	τ_{max} [min]	OIT [min]	τ_{max} [min]
CP_PSO	59.89 ^{aD} ± 0.76	61.72 ^{aD} ± 0.90	20.28 ^{aC} ± 0.08	20.96 ^{aC} ± 0.01	7.46 ^{aB} ± 0.04	7.81 ^{aB} ± 0.03	3.37 ^{aA} ± 0.26	3.75 ^{aA} ± 0.03
SE_PSO	104.60 ^{bD} ± 1.03	105.96 ^{bD} ± 0.83	39.13 ^{bC} ± 0.30	39.91 ^{bC} ± 0.30	14.57 ^{bB} ± 0.07	15.06 ^{bB} ± 0.10	4.67 ^{bA} ± 0.17	5.28 ^{bA} ± 0.27
UAE_PSO	105.11 ^{bD} ± 0.33	106.80 ^{bD} ± 0.64	40.00 ^{bC} ± 0.42	40.75 ^{bC} ± 0.39	14.66 ^{bB} ± 0.59	15.25 ^{bB} ± 0.40	5.18 ^{bA} ± 0.07	5.44 ^{bA} ± 0.02

CP_PSO—cold-pressed pomegranate seed oil; SE_PSO—pomegranate seed oil extracted by the Soxhlet method; UAE_PSO—pomegranate seed oil extracted by ultrasound-assisted extraction. The different lower-case letters indicate significantly different values ($p \leq 0.05$). Data are presented as mean values followed by standard deviation (±SD). Different lower-case letters (a,b) in the columns indicate statistically significant differences between different samples oxidized at the same temperature but for different obtaining methods, at the level of $\alpha = 0.05$. Different capital letters (A–D) in the rows indicate statistically significant differences between different samples obtained by the same method but at different temperatures, at the level of $\alpha = 0.05$ (separately for OIT and τ_{max}).

2.5.2. Oxidation Kinetics

The oxidation process in accelerated methods takes place at high temperatures and has a different course than at ambient temperature. However, determining the stability of oils at different temperatures allows for determining the parameters of the oxidation kinetics, which helps estimate the oil's durability [46]. The scientific literature still lacks comprehensive data concerning the oxidation kinetics of PSO.

The oxidation of oils is a multi-step reaction that typically proceeds through a free radical mechanism. According to Kowalski et al. (2004), the oxidation of seed oils in excess of oxygen follows a first-order reaction mechanism [51]. However, it should be noted that oxidation of oils is theoretically not a first-order reaction, but such a model can be used as an approximation in kinetic analysis. Kaseke et al. (2021) fitted different kinetic models to evaluate lipid oxidation kinetics, and they selected the first-order one as the best fitting reaction order [52].

Determination of the kinetics of vegetable oil oxidation was primarily related to the analysis of the activation energy (E_a) and the reaction rate constant (k). The results of the calculations of kinetic parameters are listed in Table 6. The τ_{max} measured under isothermal conditions (90–120 °C) using the PDSC method allowed for the preparation of a graphical relationship between the τ_{max} logarithm and the reciprocal of temperature. R^2 coefficients above 0.99 meant that the data describing linear correlations could be used for further calculations. The E_a of the oxidation reaction for oil samples extracted from pomegranate seeds ranged from 106.15 to 111.99 kJ/mol. Kaseke et al. (2021) determined the E_a for the lipid oxidation reaction in PSO extracted from unpretreated, blanched, and microwaved seeds as equal to 6.29, 8.98, and 8.59 kJ/mol, respectively. However, the authors used a different method to study oxidation kinetics using peroxide values to assess the oxidative stability of PSO samples stored at 50 °C [52]. The E_a values depend on the measurement method used as well as the experimental conditions. In the PDSC method, pure oxygen under pressure is applied, which may lead to higher energy activation values.

Table 6. Regression analysis of the DSC data at different temperatures for pomegranate seed oils.

Parameter	CP_PSO	SE_PSO	UAE_PSO
$-a$	5.83 ± 0.10	5.96 ± 0.10	6.15 ± 0.10
b	14.29 ± 0.10	14.39 ± 0.10	14.89 ± 0.10
R^2	0.9963	0.9998	0.9986
E_a (kJ/mol)	106.15 ± 0.10	108.59 ± 0.10	111.99 ± 0.10
Z (min^{-1})	3.14×10^{12}	3.87×10^{12}	3.14×10^{12}
k at 90 °C (min^{-1})	0.0017	0.0009	0.0090
k at 100 °C (min^{-1})	0.0043	0.0024	0.0025
k at 110 °C (min^{-1})	0.0106	0.0061	0.0064
k at 120 °C (min^{-1})	0.0247	0.0145	0.0157

CP_PSO—cold-pressed pomegranate seed oil; SE_PSO—pomegranate seed oil extracted by the Soxhlet method; UAE_PSO—pomegranate seed oil extracted by ultrasound-assisted extraction. E_a—activation energy, Z—pre-exponential factor, k—oxidation reaction rate constant.

Similar to the results in the present paper, the studies by Wirkowska et al. [53] showed a strong inverse relationship between temperature and τ_{max}, which enabled the determination of kinetic parameters due to the high agreement of the linear model ($R^2 > 0.99$). Amaranth and quinoa oils were the subject of the scientists' research. It was found that the E_a of oxidation of amaranth and quinoa oils was lower than in the case of linseed, camelina, walnut, olive, and rapeseed oils, which may be related to the higher content of unsaturated FA in these oils.

The E_a of the oxidation reaction for linseed oils studied by Symoniuk et al. [54] was lower than for the samples analyzed in this study and ranged from 93.14 to 94.53 kJ/mol, while for rapeseed oil, according to Symoniuk et al. [55], E_a ranged from 86.71 to 90.54 kJ/mol. The E_a values obtained in the present study were also higher than those determined by Ciemniewska-Żytkiewicz et al. [56] for hazelnut oil (89.06 kJ/mol), rapeseed oil (92.68 kJ/mol), and olive oil (92.81 kJ/mol), and by Ratusz et al. [44], for camelina oil (87.63–93.61 kJ/mol). In turn, Farhoosh et al. [57] reported activation energy values for corn oil—88.14 kJ/mol, sunflower oil—90.24 kJ/mol, and soybean oil—92.42 kJ/mol, which were also lower than those obtained in this study.

Taking into account the literature data, E_a values exceeding 100 kJ/mol can be considered high. Despite the significant content of unsaturated FA in PSO, which promotes oxidation processes, it was observed that more energy was required to initiate the reaction than in many other oils. PSO, although it may exhibit high E_a under controlled conditions, is characterized by low oxidative stability due to the presence of highly reactive PA. Once oxidation is initiated, the process proceeds rapidly, making the oil very sensitive to oxygen, light, and heat.

The oxidative stability of vegetable oils depends not only on their FA composition, but also on the presence of natural antioxidant compounds, which can significantly slow down degradation processes. In the study conducted by Song et al. [58], the kinetic parameters of oxidation of different oils at 180 °C were compared using the Ozawa–Flynn–Wall methods and the Arrhenius equation. It was shown that the presence of natural or synthetic antioxidants increased the E_a and decreased the value of k, which resulted in a slower oxidation process. In this study, comparing different extraction methods, it was found that the oil obtained by sonication was characterized by the highest E_a value, which may indicate its greater oxidation resistance. Similar results were obtained in the study by Piasecka et al. [59], which aimed to determine the optimal conditions for the extraction of cranberry seed oil using UAE and to assess the effect of this method on the properties of the final product. Based on PDSC measurements carried out at five temperatures, the kinetic parameters of the oxidation reaction were calculated, and, similarly to this study, a higher E_a value was found for the oil extracted by the method assisted by ultrasounds than for oil obtained by the conventional solid–liquid extraction with *n*-hexane. It is worth noting that the results of Piasecka et al. [59] partially contrast with the reports of Pérez-Sauced et al. [60], who showed that the use of ultrasound for the extraction of avocado oil resulted in a lower E_a compared to cold extraction by centrifugation. These differences confirm that the extraction method and its conditions have a significant impact on the oxidative stability of the obtained oils, and the effect of ultrasound may depend on the type of raw material, process parameters, and chemical composition of the oil.

Considering the reaction rate coefficients obtained in this paper, it can be stated that they increased with increasing temperature. For example, at 90 °C, the rate coefficient value was from 0.0009 to 0.009 min^{-1}, while at 120 °C, it was from 0.0145 to 0.0247 min^{-1}. In the study by Symoniuk et al. [54] it was shown that *k* for linseed oil, calculated based on the results obtained by the PDSC method at 100 °C, were higher than the values presented by Ciemniewska-Żytkiewicz et al. [56] for hazelnut oil (0.0071 min^{-1}), olive oil (0.0066 min^{-1}), and rapeseed oil (0.0108 min^{-1}), and ranged from 0.0636 to 0.0890 min^{-1}. These results confirm the greater susceptibility of linseed oil to oxidation. For comparison, *k* for the tested PSO samples at the same temperature (100 °C) ranged from 0.0024 to 0.0043 min^{-1}, which indicates their moderate susceptibility to oxidation.

2.6. Bioactive Properties

PSO is a valuable source of numerous bioactive compounds that confer significant health-promoting properties. Particular attention is given to its high content of conjugated FAs, primarily PA, which constitutes the dominant lipid fraction in the oil and belongs to the group of conjugated linolenic acids (CLnA). This compound has well-documented anti-cancer and anti-inflammatory activities [12,61,62]. PSO also contains a variety of phenolic compounds, which are key contributors to its free radical scavenging capacity. These phenolics support cellular defense mechanisms against oxidative stress, thereby delaying cellular aging and degeneration processes [36]. Other important bioactive constituents include tocopherols (mainly γ- and δ-tocopherol), which act as natural lipid-soluble antioxidants; phytosterols (such as β-sitosterol and campesterol), known for their hypolipidemic and anti-inflammatory effects [47]; and flavonoids, which further enhance antioxidant activity through synergistic interactions with other bioactive components present in PSO [41].

The study demonstrated that the extraction method influenced the content of bioactive compounds in PSO (Table 7). Oil extracted using the Soxhlet method contained fewer polyphenols compared to other tested oils due to the high temperature and prolonged extraction times involved in the process. Some polyphenols are heat sensitive, and they tend to degrade or remain in the seeds during solvent extraction. Additionally, the use of organic solvents may not be as selective for polyphenols compared to other, gentler extraction methods like cold pressing or UAE. Cold-pressed oil exhibited a significantly higher polyphenol content than oil extracted using the Soxhlet method and oil obtained through UAE. The polyphenol content in CP_PSO was approximately 1.7 times higher than in SE_PSO and 1.2 times higher than in UAE_PSO. These findings are consistent with previous research, which indicates that non-thermal extraction methods, such as cold pressing, more effectively preserve thermolabile polyphenolic compounds than solvent-based methods [29]. However, it should be noted that cold-pressed oil, although it had the highest polyphenol content, also had the lowest oxidative stability (Tables 5 and 7). The cold-pressing process did not include refining steps that could remove pro-oxidative compounds. These steps help eliminate free fatty acids, phospholipids, pigments, minerals, metals, and other pro-oxidative substances that can accelerate oxidation. Although refining also reduces the content of natural antioxidants, the removal of oxidation-promoting compounds often results in a more oxidatively stable oil compared to cold-pressed oil. Unrefined cold-pressed oils are more vulnerable to degradation when exposed to air, light, or heat—even though they contain a high level of antioxidants.

Table 7. Total polyphenol content (TPC), ABTS, and DPPH radical scavenging capacity of obtained pomegranate seed oils.

Oil	TPC [mg GAE/1 g of Oil]	Antioxidant Activity ABTS [μmol TE/100 g]	Antioxidant Activity DPPH [μmol TE/100 g]
CP_PSO	$2.52^{c} \pm 0.10$	$445.04^{c} \pm 2.12$	$1666.19^{c} \pm 6.33$
SE_PSO	$1.48^{a} \pm 0.11$	$401.65^{a} \pm 2.83$	$1482.94^{a} \pm 4.88$
UAE_PSO	$2.08^{b} \pm 0.14$	$418.69^{b} \pm 3.29$	$1508.27^{b} \pm 5.39$

CP_PSO—cold-pressed pomegranate seed oil; SE_PSO—pomegranate seed oil extracted by the Soxhlet method; UAE_PSO—pomegranate seed oil extracted by ultrasound-assisted extraction. The different lower-case letters show a statistically significant difference in the column at a $p \leq 0.05$. Data are presented as mean values followed by standard deviation (±SD).

In the study by Khoddami et al. [63], oil from three different pomegranate seed varieties was cold-pressed, and the reported total polyphenol content (TPC) ranged from 8.52 to 10.44 mg GAE/g PSO, which is approximately four times higher than the values obtained in the present study. In contrast, in the study by Abbasi et al. [64], PSO extracted using the Soxhlet method was characterized by TPC of 9 mg GAE/g oil. The study by Juhaimi et al. [36] demonstrated that Soxhlet extraction yielded oil with TPC ranging from 7.8 to 19.2 mg GAE/g. TPC of pomegranate seed oil from the Hicaz variety was found to be 8.71 mg GAE/g [36], which is significantly higher than the value obtained in this study. On the other hand, the TPC values observed in this study are significantly higher than those reported for PSO from the Tounsi variety (93.42 mg GAE/kg) [65].

Antioxidant activity determined using the ABTS method indicates that cold-pressed oil exhibits the highest capacity for neutralizing free radicals. A similar trend was observed in antioxidant activity measured using the DPPH method.

Cold-pressed pomegranate seed oil [29] showed markedly higher ABTS and DPPH radical scavenging activity compared to the oil examined in the present study, indicating substantial differences in their antioxidant potential. In the study by Rojo-Gutiérrez et al. [66], the ABTS and DPPH radical scavenging activity of PSO obtained using UAE was significantly higher than the values obtained in this work. Few available scientific publications focus specifically on the determination of the antioxidant activity of PSO. In the majority of studies concerning pomegranate seeds, antioxidant activity values measured by ABTS and DPPH assays are reported in relation to ethanolic, acetonic, or aqueous extracts, or expressed per unit of dry seed mass [67–69]. Consequently, there is a need for further research aimed at standardizing analytical methodologies specific to PSO, to enable reliable comparisons across different sample types and extraction methods.

Differences in TPC and antioxidant activity of PSO between various extraction methods can be attributed to variations in solvent selectivity, as well as the applied temperature and extraction duration. The higher TPC and antioxidant potential of cold-pressed oil may be due to the absence of heat exposure, which protects phenolic compounds from degradation. Conversely, the lower efficiency of Soxhlet and UAE extraction is likely a result of thermal and solvent-induced degradation of bioactive compounds [70,71]. Additionally, the observed differences may arise from the varying ability of each method to extract hydrophilic and lipophilic bioactive substances—while the ABTS method measures both hydrophilic and lipophilic antioxidants, the DPPH method primarily detects lipophilic compounds [72,73]. Furthermore, the content of bioactive compounds and antioxidant activity in fruit seed oils is influenced by factors such as origin, variety, ripening stage, and seed quality [74]. Storage and transportation conditions of the seeds may also play a crucial role. It is important to note that the analyzed oils were extracted from seeds that were a by-product of juice pressing rather than directly from fresh fruit seeds.

3. Materials and Methods

3.1. Materials

The pomegranate seed oils were obtained from raw materials (*Punica granatum* L. var. Hicaz) grown in Croatia. After juice production, the pomegranate seeds were dried in the sun and transported to Poland in airtight, sterile thermal containers to protect them from moisture and light. The Folin–Ciocalteu reagent, DPPH reagent, ABTS reagent, and *p*-anisidine were purchased from Sigma-Aldrich (USA), and other chemicals were purchased from Chempur (Poland).

3.2. Methods

3.2.1. Oil Obtaining Methods

Cold Pressing

Oil was collected from 200 g of pomegranate seeds using an automated oil cold-pressing machine YODA YD-ZY-02A (Yoda, Warsaw, Poland). The temperature program was set in the range of 32–38 °C. The pressing time was around 8 min. The oil was carried down to a decantation vessel. Extraction was performed three times. The oil was collected, separated by centrifugation, and stored at 4 °C before testing.

Soxhlet Extraction

To obtain the oil by hexane extraction using a Soxhlet apparatus, 20 g (±0.001 g) of seeds was weighed and ground in an IKA Tube Mill (IKA Works GmbH & Co. KG, Staufen, Germany) at 25,000 rpm. The seeds crushed in this way were wrapped in filter paper, put into a thimble, and placed in a Soxhlet apparatus. Hexane (200 mL) was added to the round-bottom flask. The extraction was carried out for 6 h. The obtained extracts were dried with anhydrous magnesium sulfate, which was then filtered off. The solvent was distilled using a vacuum rotary evaporator Rotavapor® R-300 (BUCHI, Uster, Switzerland). Residual *n*-hexane was removed from oil samples in nitrogen atmosphere.

Experimental Design—Ultrasound-Assisted Extraction

Based on previous PSO optimization studies [19,20,26], the variables were chosen for a three-factor, three-level Box–Behnken design of experiment: ultrasound amplitude in a range of 30–90%, sonication time in a range of 6–12 min, and L/S ratio in a range of 10–20. Detailed variables of all performed runs are shown in Table 8. The responses chosen for the experiment were oil yield (%) and maximum oxidation time (τ_{max}). Then, suitable models were fit according to ANOVA results. Specific equations describing models were also determined (3- for the linear model and 4- for the quadratic model):

$$Y = \beta_0 + \beta_1 X_1 + \beta_2 X_2 + \beta_3 X_3 \quad (3)$$

$$Y = \beta_0 + \beta_1 X_1 + \beta_2 X_2 + \beta_3 X_3 + \beta_{12} X_1 X_2 + \beta_{13} X_1 X_3 + \beta_{23} X_2 X_3 + \beta_{11} {X_1}^2 + \beta_{22} {X_2}^2 + \beta_{33} {X_3}^2 \quad (4)$$

where β_0—the constant coefficient; β_1, β_2, β_3 —regression coefficients for the linear terms; β_{11}, β_{22}, β_{33}—regression coefficients for the quadratic terms; β_{12}, β_{23}, β_{13}—regression coefficients for the interaction terms; X_1, X_2, X_3—coded values of independent variables.

Table 8. The Box–Behnken design variables for ultrasound-assisted extraction of pomegranate seed oil.

Run	X_1-Amplitude (%)	X_2-Time (minute)	X_3-L/S Ratio
1	30	6	15
2	30	12	15
3	90	6	15
4	90	12	15
5	60	6	10
6	60	12	10
7	60	6	20
8	60	12	20
9	30	9	10
10	90	9	10
11	30	9	20
12	90	9	20
13	60	9	15
14	60	9	15
15	60	9	15

Ultrasound-Assisted Extraction

UAE of the oils was carried out in a UP200Ht ultrasonic processor supplied by Hielscher Ultrasonics GmbH (Tetlow, Germany) with an output power of 200 W, according to the method proposed by Piasecka et al. [75] with minor modifications. Seed samples (5 ± 0.001 g) were ground using an IKA tube mill (IKA-Werke, Staufen im Breisgau, Germany) at 20,000 rpm for 60 s, then placed in a Falcon tube and filled with *n*-hexane at a proper L/S ratio just before extraction. The resulting extracts were filtered, dried with anhydrous magnesium sulfate, and evaporated at 70 mbar. The remaining *n*-hexane was removed under a nitrogen atmosphere. The parameters chosen as optimal process conditions were 46% amplitude, 12 min, and 19 L/S ratio.

3.2.2. Oil Yield Determination

The oil yield obtained from pomegranate seeds was measured using a gravimetric test, calculated by dividing the oil's weight by the seeds' initial weight and expressed as a percentage, according to Equation (5) [76].

$$Yield\ [\%] = \frac{m_o}{m_s} \times 100 \quad (5)$$

where m_o—mass of oil; m_s—mass of seeds.

3.2.3. Determination of Fatty Acid Composition by Gas Chromatography

A gas chromatograph YL6100 GC (Young Lin Bldg., Anyang, Hogyedong, Republic of Korea) equipped with a BPX-70 capillary column (SGE Analytical Science, Milton Keynes, UK) with the following parameters: length: 60 m; film thickness: 0.25 μm; and inner diameter: 0.25 mm; and a flame ionization detector was applied to determine the composition of fatty acids present in the oils. The fatty acid methyl esters (FAME) were prepared according to PN-EN ISO 5509:2001 [77]. The gas that carries FAME inside the column was nitrogen. The injector and detector temperatures were set at 225 °C and 250 °C, respectively. The temperature program for the GC started at 70 °C, held for 0.5 min, then increased at (I) 15 °C/min to 160 °C; (II) 1.1 °C/min from 160 °C to 200 °C (held for 12 min) and (III) 30 °C/min from 200 °C to 225 °C.

The individual FAs were identified by comparing the retention times of FAME. The abundance percentage of each FA was also determined. The determinations were performed at least in duplicate.

3.2.4. Determination of Acid, Peroxide, and p-Anisidine Values and TOTOX Index

The degree of hydrolysis of the tested oils was determined using AV according to the AOCS Cd 3d-63 [78]. The content of primary oxidation products of the oils was examined by PV according to the AOCS method (AOCS Cd 8–53) [79]. The degree of secondary oxidation of oils was determined by p-AnV according to the AOCS method (AOCS Cd 18–90) [80]. The TOTOX index, indicating the total oxidation value, was calculated according to the formula: TOTOX index = (2 × PV) + p-AnV.

3.2.5. Determination of Oxidative Stability and the Oxidation Kinetics Using the PDSC Method

Pressure differential scanning calorimetry using DSC Q20 TA Instruments (Newcastle, WA, USA) was employed to determine the OIT and $\tau_{\max}$ as well as kinetic parameters for the oxidation reaction of oils. The weight of the oils used in the tests ranged from 3 to 4 mg. Oil placed in an open aluminum pan was oxidized at a pressure of ~1400 kPa in a cell with an empty reference pan. The samples were tested under isothermal conditions. The accelerated oxidation of oil samples in a PDSC cell was set at temperatures of 90, 100, 110, and 120 °C.

The OIT was determined as the intersection of the extrapolated baseline and the tangent line (leading edge) of the exothermic curve and was measured in minutes, while $\tau_{\max}$ was determined as the time corresponding to the maximum of the PDSC signal.

The isothermal PDSC variant was employed to assess oxidation kinetic parameters by using the following Equations (6)–(8) [59] and according to the Ozawa–Flynn–Wall methodology:

$$log\tau_{max} = a\cdot\left(\frac{1}{T}\right) + b \quad (6)$$

where τ_{max} is maximum oxidation time (min) and T is the temperature (K);

$$E_a = 2.19\cdot R\cdot a \quad (7)$$

where E_a—activation energy, R ($J\cdot mol^{-1}\cdot K^{-1}$) is a gas constant, and a is a coefficient from Equation (6),

$$k = Ze^{\frac{-E_a}{RT}} \quad (8)$$

where k is a reaction rate coefficient (min^{-1}); Z is a pre-exponential factor.

3.2.6. Preparation of Extracts for Analysis of Bioactive Properties of Oils

A total of 0.5 g of oil was mixed with 2.5 mL of hexane and 2.5 mL of methanol. The mixture was then shaken for 2 min, centrifuged (4000 rpm) for 10 min, and the bottom layer was collected in a new tube. The procedure was repeated 3 times.

3.2.7. Determination of Total Polyphenol Content (TPC)

The method of determination proposed by Gao et al. [81] was used to determine the content of total polyphenols. In test tubes, 200 μL of sample, 0.3 mL of Folin–Ciocalteu reagent, as well as 4.9 mL of distilled water and 0.6 mL of 20% sodium carbonate solution were measured sequentially. The contents of the tubes were mixed using a Vortex centrifuge and left in a dark place for 60 min. After this time, the absorbance at λ = 765 nm was measured against the blank, which was methanol, using a Jenway 6305 UV–VIS spectrophotometer (Cole-Parmer, Vernon Hills, IL, USA). The calibration curve was prepared using working standard solutions of gallic acid at concentrations ranging from 100 mg to 1200 mg/L. TPC values were given in milligrams of gallic acid equivalent (GAE) per 1 g of oil.

3.2.8. Determination of Antioxidant Activity Using ABTS Cation Radicals

To determine the antioxidant activity of the extracts obtained from the oils, the method described by Re et al. [82] with modifications was used. For this purpose, an ABTS (2,2′-azino-bis-(3-ethylbenzothiazoline-6-sulfonic acid) solution was prepared by mixing 14 mM ABTS solution with 4.9 mM of potassium persulfate solution in a 1:1 ratio. The prepared solution was left for a minimum of 16 h in a place inaccessible to light, at 4 °C. The solution was then diluted so that the absorbance of the cation radical solution measured at λ = 734 nm was in the range of 0.680–0.720. A Trolox standard curve was created using working standard solutions at concentrations ranging from 0 to 1125 µmol/L. Antioxidant capacity was given as µmol Trolox equivalent (TE) per 100 g of oil.

3.2.9. Determination of Antioxidant Activity Using DPPH Cation Radicals

Antioxidant activity was also measured using the method described by Brand-Williams et al. [83] with modifications. For this purpose, a 0.6 mM solution of DPPH (1,1-diphenyl-2-picrylhydrazyl) was prepared and left for about 16 h in a dark place at 4 °C. The solution was diluted so that the absorbance measured at λ = 515 nm was in the range of 0.680–0.720. Then, 100 µL of the sample was measured into test tubes, and 4 mL of DPPH radical solution was added. The contents of the tubes were mixed using a Vortex centrifuge, and after 30 min of incubation in the dark, the absorbance against methanol as a blank was measured at λ = 515 nm. A Trolox standard curve was created as for ABTS antioxidant capacity assay. The result was given as µmol Trolox equivalent (TE) per 100 g of oil.

3.2.10. Statistical Analysis

Statistical software Statistica 13 was used to statistically process the results obtained. One-factor analysis of variance (ANOVA) and Tukey's post hoc test at the significance level of α = 0.05 were used for the following determinations: bioactive properties, quality parameters of oils, fatty acid profile of oils, and yield, while two-factor ANOVA was used for the results on the oxidative stability of the oils obtained. Microsoft Excel 2010 was used to calculate the mean and standard deviation. All analyses were carried out in triplicate for each sample tested.

4. Conclusions

The results of this study demonstrate that the method of extraction significantly affects the yield, oxidative stability, quality parameters, and bioactive compounds content of PSO. UAE proved to be the most efficient technique in terms of oil yield, achieving 12.67% while maintaining favorable oxidative stability and moderate levels of bioactive compounds. Cold pressing, although less efficient in terms of yield, preserved the highest concentration of polyphenols and exhibited the strongest antioxidant activity, but showed the lowest oxidative stability and the highest susceptibility to lipid oxidation. Soxhlet extraction produced oil with the highest content of punicic acid but with exceeded recommended acid value limits, indicating possible hydrolytic degradation during extraction. The fitted statistical models, based on the Box–Behnken design, effectively predicted the extraction yield and oxidation parameters, confirming the reliability of the optimization process. Compared to conventional methods, UAE presents a promising alternative, balancing high efficiency with minimal degradation of thermolabile compounds and reduced environmental impact, aligning with circular economy and sustainability principles.

Future research should focus on quantifying individual bioactive compounds that may significantly influence the kinetics of lipid oxidation by altering reaction rates, activation energy, and even the underlying mechanism, depending on their concentration and stability. While some act as antioxidants, slowing the reaction, others may exhibit pro-oxidant effects under certain conditions. These findings highlight the need for more comprehensive studies to fully understand their role in the oxidative stability of pomegranate seed oil.

Author Contributions: Conceptualization, M.S.; methodology, M.S., I.P. and J.B.; software, J.B. and A.G.; validation, M.S., I.P. and J.B.; formal analysis, M.S.; investigation, M.S. and I.P.; resources, J.B. and A.G.; data curation, M.S.; writing—original draft preparation, M.S., I.P., D.M.-J. and J.B.; writing—review and editing, M.S., I.P., D.M.-J., A.G. and J.B.; visualization, M.S.; supervision, J.B., D.M.-J. and A.G.; project administration, M.S.; funding acquisition, M.S. All authors have read and agreed to the published version of the manuscript.

Funding: This research received no external funding.

Institutional Review Board Statement: Not applicable.

Informed Consent Statement: Not applicable.

Data Availability Statement: Data are contained within the article.

Acknowledgments: The research for this publication was carried out with the use of equipment purchased as part of the "Food and Nutrition Centre—modernisation of the WULS campus to create a Food and Nutrition Research and Development Centre (CŻiŻ)" co-financed by the European Union from the European Regional Development Fund under the Regional Operational Programme of the Mazowieckie Voivodeship for 2014–2020 (Project No. RPMA.01.01.00-14-8276/17). Special acknowledgements go to the Novallis company, which provided the material for the present study.

Conflicts of Interest: The authors declare no conflicts of interest.

Abbreviations

The following abbreviations are used in this manuscript:

PSO	Pomegranate seed oil
PA	Punicic acid
FA	Fatty acid
CP_PSO	Cold-pressed pomegranate seed oil
SE_PSO	Pomegranate seed oil extracted by the Soxhlet method
UAE_PSO	Pomegranate seed oil extracted by the ultrasound-assisted method
UAE	Ultrasound-assisted extraction
GC	Gas chromatography
PDSC	Pressure differential scanning calorimetry
FAME	Fatty acid methyl esters
OIT	Oxidation induction time
L/S	Liquid to solid
TAG	Triacylglycerol
PUFA	Polyunsaturated fatty acid
MUFA	Monounsaturated fatty acid
SFA	Saturated fatty acid
AV	Acid value
PV	Peroxide value
p—AnV	Anisidine value
TPC	Total polyphenol content

CLnA	Conjugated linolenic acids
SHHE	Superheated hexane extraction

References

1. Chandrasekar, V.; Liming, Z.; Ruihong, Z.; Zhongli, P. Chapter 8—Pomegranate. In *Integrated Processing Technologies for Food and Agricultural By-Products*; Pan, Z., Zhang, R., Zicari, S., Eds.; Academic Press: New York, NY, USA, 2019; pp. 181–216. [CrossRef]
2. Wang, J.; Sun, M.; Yu, J.; Wang, J.; Cui, Q. Pomegranate seeds: A comprehensive review of traditional uses, chemical composition, and pharmacological properties. *Front. Pharmacol.* **2024**, *15*, 1401826. [CrossRef]
3. Fidelis, M.; de Moura, C.; Kabbas Junior, T.; Pap, N.; Mattila, P.; Mäkinen, S.; Putnik, P.; Bursać Kovačević, D.; Tian, Y.; Yang, B.; et al. Fruit Seeds as Sources of Bioactive Compounds: Sustainable Production of High Value-Added Ingredients from By-Products within Circular Economy. *Molecules* **2019**, *24*, 3854. [CrossRef]
4. Kowalska, H.; Czajkowska, K.; Cichowska, J.; Lenart, A. What's new in biopotential of fruit and vegetable by-products applied in the food processing industry. *Trends Food Sci. Technol.* **2017**, *67*, 150–159. [CrossRef]
5. Lau, K.Q.; Sabran, M.R.; Shafie, S.R. Utilization of vegetable and fruit by-products as functional ingredient and food. *Front. Nutr.* **2021**, *8*, 661693. [CrossRef] [PubMed]
6. Ko, K.; Dadmohammadi, Y.; Abbaspourrad, A. Nutritional and Bioactive Components of Pomegranate Waste Used in Food and Cosmetic Applications: A Review. *Foods* **2021**, *10*, 657. [CrossRef]
7. Sreekumar, S.; Sithul, H.; Muraleedharan, P.; Azeez, J.M.; Sreeharshan, S. Pomegranate Fruit as a Rich Source of Biologically Active Compounds. *BioMed Res. Int.* **2014**, *2014*, 686921. [CrossRef] [PubMed]
8. de Melo, I.L.P.; de Carvalho, E.B.T.; e Silva, A.M.d.O.; Yoshime, L.T.; Sattler, J.A.G.; Pavan, R.T.; Mancini-Filho, J. Characterization of Constituents, Quality and Stability of Pomegranate Seed Oil (*Punica granatum* L.). *Food Sci. Technol.* **2016**, *36*, 132–139. [CrossRef]
9. Verardo, V.; Garcia-Salas, P.; Baldi, E.; Segura-Carretero, A.; Fernandez-Gutierrez, A.; Caboni, M.F. Pomegranate Seeds as a Source of Nutraceutical Oil Naturally Rich in Bioactive Lipids. *Food Res. Int.* **2014**, *65*, 445–452. [CrossRef]
10. Quitmeyer, B.; Emelife, C.; Klausner, H.; Gbayisomore, O.; Phelan, S. Differential Effects of Punicic Acid on Cytotoxicity and Peroxiredoxin Expression in MCF-7 Breast Cancer and MCF-10A Normal Cells. *Anticancer Res.* **2024**, *44*, 4751–4759. [CrossRef]
11. Chen, L.; Lei, Y.; Lu, C.; Liu, D.; Ma, W.; Lu, H.; Wang, Y. Punicic Acid Ameliorates Obesity-Related Hyperlipidemia and Fatty Liver in Mice via Regulation of Intestinal Flora and Lipopolysaccharide-Related Signaling Pathways. *Food Funct.* **2024**, *15*, 5012–5025. [CrossRef]
12. Aruna, P.; Venkataramanamma, D.; Singh, A.K.; Singh, R.P. Health Benefits of Punicic Acid: A Review. *Comp. Rev. Food Sci. Food Safe* **2016**, *15*, 16–27. [CrossRef] [PubMed]
13. Vroegrijk, I.O.C.M.; van Diepen, J.A.; van den Berg, S.; Westbroek, I.; Keizer, H.; Gambelli, L.; Hontecillas, R.; Bassaganya-Riera, J.; Zondag, G.C.M.; Romijn, J.A.; et al. Pomegranate Seed Oil, a Rich Source of Punicic Acid, Prevents Diet-Induced Obesity and Insulin Resistance in Mice. *Food Chem. Toxicol.* **2011**, *49*, 1426–1430. [CrossRef] [PubMed]
14. Khajebishak, Y.; Payahoo, L.; Alivand, M.; Alipour, B. Punicic Acid: A Potential Compound of Pomegranate Seed Oil in Type 2 Diabetes Mellitus Management. *J. Cell Physiol.* **2019**, *234*, 2112–2120. [CrossRef]
15. Seher, S.S.; Ali, N.A.; Seher, S.S.; Ali, N.A. Pomegranate Seeds: The Antioxidant Marvel and Its Therapeutic Journey. In *Advances in Medical Diagnosis, Treatment, and Care*; Musaddiq, S., Fayyaz, I., Mustafa, K., Eds.; IGI Global: Hershey, PA, USA, 2024; pp. 189–240, ISBN 9798369319864.
16. Badr, A.N.; Ali, H.S.; Abdel-Razek, A.G.; Shehata, M.G.; Albaridi, N.A. Bioactive Components of Pomegranate Oil and Their Influence on Mycotoxin Secretion. *Toxins* **2020**, *12*, 748. [CrossRef]
17. Ćurko, N.; Lukić, K.; Tušek, A.J.; Balbino, S.; Vukušić Pavičić, T.; Tomašević, M.; Redovniković, I.R.; Ganić, K.K. Effect of Cold Pressing and Supercritical CO_2 Extraction Assisted with Pulsed Electric Fields Pretreatment on Grape Seed Oil Yield, Composition and Antioxidant Characteristics. *LWT* **2023**, *184*, 114974. [CrossRef]
18. Prescha, A.; Grajzer, M.; Dedyk, M.; Grajeta, H. The Antioxidant Activity and Oxidative Stability of Cold-Pressed Oils. *J. Am. Oil Chem. Soc.* **2014**, *91*, 1291–1301. [CrossRef] [PubMed]
19. Hawthorne, S.B.; Grabanski, C.B.; Martin, E.; Miller, D.J. Comparisons of Soxhlet Extraction, Pressurized Liquid Extraction, Supercritical Fluid Extraction and Subcritical Water Extraction for Environmental Solids: Recovery, Selectivity and Effects on Sample Matrix. *J. Chromatogr. A* **2000**, *892*, 421–433. [CrossRef]
20. Saputra, H.; Azrifirwan, A.; Firdani, F. Evaluation of Vegetable Oil Extraction Methods on Crude Oil Yields in Indonesia. Systematic Literature Review. *GreenTech* **2024**, *1*, 106–115. [CrossRef]
21. Staicu, V.; Tomescu, J.A.; Neagu, M.; Popescu, A.F.; Popescu, M.; Manea, C.; Călinescu, I. Ultrasonic-Assisted Extraction of Bioactive Compounds from Elecampagne (*Inula helenium*, Radix Inulae). *J. Biomed. Res. Env. Sci.* **2023**, *4*, 1184–1192. [CrossRef]

22. Khalid, S.; Chaudhary, K.; Amin, S.; Raana, S.; Zahid, M.; Naeem, M.; Mousavi Khaneghah, A.; Aadil, R.M. Recent Advances in the Implementation of Ultrasound Technology for the Extraction of Essential Oils from Terrestrial Plant Materials: A Comprehensive Review. *Ultrason. Sonochem.* **2024**, *107*, 106914. [CrossRef]
23. Zhao, Y.; Guo, S.; Li, S.; Ye, E.; Wang, W.; Wang, T.; Wen, Y.; Guo, L. Ultrasonic-Assisted Extraction, Anti-Biofilm Activity, and Mechanism of Action of Ku Shen (*Sophorae Flavescentis Radix*) Extracts against *Vibrio parahaemolyticus*. *Front. Microbiol.* **2024**, *15*, 1379341. [CrossRef]
24. Tian, Y.; Xu, Z.; Zheng, B.; Martin Lo, Y. Optimization of Ultrasonic-Assisted Extraction of Pomegranate (*Punica granatum* L.) Seed Oil. *Ultrason. Sonochem.* **2013**, *20*, 202–208. [CrossRef]
25. Barizão, É.O.; Boeing, J.S.; Martins, A.C.; Visentainer, J.V.; Almeida, V.C. Application of Response Surface Methodology for the Optimization of Ultrasound-Assisted Extraction of Pomegranate (*Punica granatum* L.) Seed Oil. *Food Anal. Methods* **2015**, *8*, 2392–2400. [CrossRef]
26. Paul, A.; Radhakrishnan, M. Pomegranate Seed Oil in Food Industry: Extraction, Characterization, and Applications. *Trends Food Sci. Technol.* **2020**, *105*, 273–283. [CrossRef]
27. Hosseini, S.; Gharachorloo, M.; Tarzi, B.G.; Ghavami, M.; Bakhoda, H. Effects of Ultrasound Amplitude on the Physicochemical Properties of Some Edible Oils. *J. Am. Oil Chem. Soc.* **2015**, *92*, 1717–1724. [CrossRef]
28. Dib, S.; Mostefa-Kara, B.; Villemin, D. Extraction and Composition of Lipids from Pomegranate Seed Oil from West Algeria. *Chem. Proc.* **2023**, *14*, 78. [CrossRef]
29. Kaseke, T.; Opara, U.L.; Fawole, O.A. Quality and Antioxidant Properties of Cold-Pressed Oil from Blanched and Microwave-Pretreated Pomegranate Seed. *Foods* **2021**, *10*, 712. [CrossRef] [PubMed]
30. Hernández-Santos, B.; Rodríguez-Miranda, J.; Herman-Lara, E.; Torruco-Uco, J.G.; Carmona-García, R.; Juárez-Barrientos, J.M.; Chávez-Zamudio, R.; Martínez-Sánchez, C.E. Effect of oil extraction assisted by ultrasound on the physicochemical properties and fatty acid profile of pumpkin seed oil (*Cucurbita pepo*). *Ultrason. Sonochem.* **2016**, *31*, 429–436. [CrossRef]
31. Thilakarathna, R.C.N.; Siow, L.F.; Tang, T.-K.; Lee, Y.Y. A review on application of ultrasound and ultrasound assisted technology for seed oil extraction. *J. Food Sci. Technol.* **2023**, *60*, 1222–1236. [CrossRef]
32. Boroushaki, M.T.; Mollazadeh, H.; Afshari, A.R. Pomegranate seed oil: A comprehensive review on its therapeutic effects. *Int. J. Pharm. Sci. Res.* **2016**, *7*, 430. [CrossRef]
33. Rowayshed, G.; Salama, A.; Abul-Fadl, M.; Akila-Hamza, S.; Emad, A.M. Nutritional and Chemical Evaluation for Pomegranate (*Punica granatum* L.) Fruit Peel and Seeds Powders By Products. *Middle East J. Appl. Sci.* **2013**, *3*, 169–179.
34. Habibnia, M.; Ghavami, M.; Ansaripour, M.; Vosough, S. 2012: Chemical evaluation of oils extracted from five different varieties of Iranian pomegranate seeds. *J. Food Biosci. Technol.* **2012**, *2*, 35–40.
35. Dadashi, S.; Mousazadeh, M.; Emam-Djomeh, Z.; Mousavi, S.M. Pomegranate (*Punica granatum* L.) seed: A comparative study on biochemical composition and oil physicochemical characteristics. *Int. J. Adv. Biol. Biomed. Res.* **2013**, *1*, 351–363.
36. Juhaimi, F.A.; Özcan, M.M.; Ghafoor, K. Characterization of pomegranate (*Punica granatum* L.) seed and oils. *Eur. J. Lipid Sci. Technol.* **2017**, *119*, 1700074. [CrossRef]
37. Eikani, M.H.; Golmohammad, F.; Homami, S.S. Extraction of Pomegranate (*Punica granatum* L.) Seed Oil Using Superheated Hexane. *Food Bioprod. Process.* **2012**, *90*, 32–36. [CrossRef]
38. *Codex-ALINORM 09/32/17*; Codex Alimentarius 2009. Codex Standard for Named Vegetable Oils. Codex Alimentarius Commission: Rome, Italy, 2009.
39. *PN-EN ISO 3960:2017-03*; Vegetable and Animal Oils and Fats. Determination of Peroxide Number (Reference Method). Polish Committee for Standardization: Warsaw, Poland, 2017.
40. Khoddami, A.; Roberts, T.H. Pomegranate Oil as a Valuable Pharmaceutical and Nutraceutical. *Lipid Technol.* **2015**, *27*, 40–42. [CrossRef]
41. Costa, A.M.M.; Silva, L.O.; Torres, A.G. Chemical Composition of Commercial Cold-Pressed Pomegranate (*Punica Granatum*) Seed Oil from Turkey and Israel, and the Use of Bioactive Compounds for Samples' Origin Preliminary Discrimination. *J. Food Compos. Anal.* **2019**, *75*, 8–16. [CrossRef]
42. Ghorbanzadeh, R.; Rezaei, K. Optimization of an Aqueous Extraction Process for Pomegranate Seed Oil. *J. Americ. Oil Chem. Soc.* **2017**, *94*, 1491–1501. [CrossRef]
43. Alasalvar, H.; Basyigit, B.; Turker, D.A.; Icyer, N.C.; Cam, M. Pomegranate seed oil: Extraction, shelf life prediction, and microencapsulation. *Carpathian J. Food Sci.* **2022**, *14*, 89–102. [CrossRef]
44. Ratusz, K.; Popis, E.; Ciemniewska-Żytkiewicz, H.; Wroniak, M. Oxidative stability of camelina (*Camelina sativa* L.) oil using pressure differential scanning calorimetry and Rancimat method. *J. Therm. Anal. Calorim.* **2016**, *126*, 343–351. [CrossRef]
45. Siol, M.; Dudek, A.; Bryś, J.; Mańko-Jurkowska, D.; Gruczyńska-Sękowska, E.; Makouie, S.; Palani, B.K.; Obranović, M.; Koczoń, P. Chromatographic and Thermal Characteristics, and Hydrolytic and Oxidative Stability of Commercial Pomegranate Seed Oil. *Foods* **2024**, *13*, 1370. [CrossRef] [PubMed]

46. Symoniuk, E.; Wroniak, M.; Napiórkowska, K.; Brzezińska, R.; Ratusz, K. Oxidative Stability and Antioxidant Activity of Selected Cold-Pressed Oils and Oils Mixtures. *Foods* **2022**, *11*, 1597. [CrossRef]
47. Yoshime, L.T.; de Melo, I.L.P.; Sattler, J.A.G.; Torres, R.P.; Mancini-Filho, J. Bioactive Compounds and the Antioxidant Capacities of Seed Oils from Pomegranate (*Punica granatum* L.) and Bitter Gourd (Momordica charantia L.). *Food Sci. Technol.* **2018**, *39*, 571–580. [CrossRef]
48. Symoniuk, E.; Ksibi, N.; Wroniak, M.; Lefek, M.; Ratusz, K. Oxidative Stability Analysis of Selected Oils from Unconventional Raw Materials Using Rancimat Apparatus. *Appl. Sci.* **2022**, *12*, 10355. [CrossRef]
49. Ratusz, K.; Kowalski, B.; Bekas, W.; Wirkowska, M. Monitorowanie autooksydacji oleju rzepakowego i slonecznikowego. *Rośliny Oleiste—Oilseed Crops* **2005**, *26*, 211–220.
50. Siol, M.; Witkowska, B.; Mańko-Jurkowska, D.; Makouie, S.; Bryś, J. Comprehensive Evaluation of the Nutritional Quality of Stored Watermelon Seed Oils. *Appl. Sci.* **2025**, *15*, 830. [CrossRef]
51. Kowalski, B.; Ratusz, K.; Kowalska, D.; Bekas, W. Determination of the oxidative stability of vegetable oils by differential scanning calorimetry and Rancimat measurements. *Eur. J. Lipid Sci. Technol.* **2004**, *106*, 165–169. [CrossRef]
52. Kaseke, T.; Opara, U.L.; Fawole, O.A. Oxidative stability of pomegranate seed oil from blanched and microwave pretreated seeds: Kinetic and thermodynamic studies under accelerated conditions. *J. Food Process. Preserv.* **2021**, *45*, e15798. [CrossRef]
53. Wirkowska-Wojdyła, M.; Ostrowska-Ligęza, E.; Górska, A.; Bryś, J. Application of Chromatographic and Thermal Methods to Study Fatty Acids Composition and Positional Distribution, Oxidation Kinetic Parameters and Melting Profile as Important Factors Characterizing Amaranth and Quinoa Oils. *Appl. Sci.* **2022**, *12*, 2166. [CrossRef]
54. Symoniuk, E.; Ratusz, K.; Krygier, K. Comparison of the Oxidative Stability of Linseed (*Linum usitatissimum* L.) Oil by Pressure Differential Scanning Calorimetry and Rancimat Measurements. *J. Food Sci. Technol.* **2016**, *53*, 3986–3995. [CrossRef]
55. Symoniuk, E.; Ratusz, K.; Krygier, K. Comparison of the Oxidative Stability of Cold-pressed Rapeseed Oil Using Pressure Differential Scanning Calorimetry and Rancimat Methods. *Eur. J. Lipid Sci. Technol.* **2017**, *119*, 1600182. [CrossRef]
56. Ciemniewska-Żytkiewicz, H.; Ratusz, K.; Bryś, J.; Reder, M.; Koczoń, P. Determination of the Oxidative Stability of Hazelnut Oils by PDSC and Rancimat Methods. *J. Therm. Anal. Calorim.* **2014**, *118*, 875–881. [CrossRef]
57. Farhoosh, R.; Niazmand, R.; Rezaei, M.; Sarabi, M. Kinetic Parameter Determination of Vegetable Oil Oxidation under Rancimat Test Conditions. *Eur. J. Lipid Sci. Technol.* **2008**, *110*, 587–592. [CrossRef]
58. Song, X.; Sui, X.; Jiang, L. Protection Function and Mechanism of Rosemary (Rosmarinus Officinalis L.) Extract on the Thermal Oxidative Stability of Vegetable Oils. *Foods* **2023**, *12*, 2177. [CrossRef]
59. Piasecka, I.; Brzezińska, R.; Ostrowska-Ligęza, E.; Wiktor, A.; Górska, A. Ultrasound-Assisted Extraction of Cranberry Seed Oil: Food Waste Valorization Approach. *Eur. Food Res. Technol.* **2023**, *249*, 2763–2775. [CrossRef]
60. Pérez-Saucedo, M.R.; Jiménez-Ruiz, E.I.; Rodríguez-Carpena, J.G.; Ragazzo-Sánchez, J.A.; Ulloa, J.A.; Ramírez-Ramírez, J.C.; Gastón-Peña, C.R.; Bautista-Rosales, P.U. Properties of the Avocado Oil Extracted Using Centrifugation and Ultrasound-Assisted Methods. *Food Sci. Biotechnol.* **2021**, *30*, 1051–1061. [CrossRef] [PubMed]
61. Gasmi, J.; Thomas Sanderson, J. Jacaric Acid and Its Octadecatrienoic Acid Geoisomers Induce Apoptosis Selectively in Cancerous Human Prostate Cells: A Mechanistic and 3-D Structure–Activity Study. *Phytomedicine* **2013**, *20*, 734–742. [CrossRef]
62. Mukherjee, C.; Bhattacharyya, S.; Ghosh, S.; Bhattacharyya, D.K. Dietary Effects of Punicic Acid on the Composition and Peroxidation of Rat Plasma Lipid. *J. Oleo Sci.* **2002**, *51*, 513–522. [CrossRef]
63. Khoddami, A.; Man, Y.B.C.; Roberts, T.H. Physico-chemical Properties and Fatty Acid Profile of Seed Oils from Pomegranate (*Punica granatum* L.) Extracted by Cold Pressing. *Euro. J. Lipid Sci. Tech.* **2014**, *116*, 553–562. [CrossRef]
64. Abbasi, H.; Rezaei, K.; Emamdjomeh, Z.; Mousavi, S.M.E. Effect of Various Extraction Conditions on the Phenolic Contents of Pomegranate Seed Oil. *Euro. J. Lipid Sci. Tech.* **2008**, *110*, 435–440. [CrossRef]
65. Amri, Z.; Lazreg-Aref, H.; Mekni, M.; El-Gharbi, S.; Dabbaghi, O.; Mechri, B.; Hammami, M. Oil Characterization and Lipids Class Composition of Pomegranate Seeds. *BioMed Res. Int.* **2017**, *2017*, 2037341. [CrossRef] [PubMed]
66. Rojo-Gutiérrez, E.; Carrasco-Molinar, O.; Tirado-Gallegos, J.M.; Levario-Gómez, A.; Chávez-González, M.L.; Baeza-Jiménez, R.; Buenrostro-Figueroa, J.J. Evaluation of Green Extraction Processes, Lipid Composition and Antioxidant Activity of Pomegranate Seed Oil. *Food Meas.* **2021**, *15*, 2098–2107. [CrossRef]
67. Jing, P.; Ye, T.; Shi, H.; Sheng, Y.; Slavin, M.; Gao, B.; Liu, L.; Yu, L. Antioxidant Properties and Phytochemical Composition of China-Grown Pomegranate Seeds. *Food Chem.* **2012**, *132*, 1457–1464. [CrossRef]
68. Peng, Y. Comparative Analysis of the Biological Components of Pomegranate Seed from Different Cultivars. *Int. J. Food Prop.* **2019**, *22*, 784–794. [CrossRef]
69. Jalal, H.; Pal, M.A.; Hamdani, H.; Rovida, M.; Khan, N.N. Antioxidant Activity of Pomegranate Peel and Seed Powder Extracts. *J. Pharmacogn. Phytochem.* **2018**, *7*, 992–997.
70. Goswami, M.J.; Dutta, U.; Kakati, D. Ultrasound-Assisted Extraction for Food, Pharmacy, and Biotech Industries. In *Bioactive Extraction and Application in Food and Nutraceutical Industries*; Sarkar, T., Pati, S., Eds.; Springer: New York, NY, USA, 2024; pp. 103–128, ISBN 9781071636015.

71. Rodrigo, S.K.; Herath, A.T. Effect of Soxhlet and Ultrasound-Assisted Solvent Extraction Methods on the Bioactivity of Adenanthera Pavonina. *S. Asian Res. J. Nat. Prod.* **2024**, *7*, 451–460. [CrossRef]
72. Christodouleas, D.C.; Fotakis, C.; Nikokavoura, A.; Papadopoulos, K.; Calokerinos, A.C. Modified DPPH and ABTS Assays to Assess the Antioxidant Profile of Untreated Oils. *Food Anal. Methods* **2015**, *8*, 1294–1302. [CrossRef]
73. Cano, A.; Alcaraz, O.; Acosta, M.; Arnao, M.B. On-Line Antioxidant Activity Determination: Comparison of Hydrophilic and Lipophilic Antioxidant Activity Using the $ABTS^{•+}$ Assay. *Redox Rep.* **2002**, *7*, 103–109. [CrossRef]
74. Gök, A.; Uyar, H.; Demir, Ö. Pomegranate Seed Oil Extraction by Cold Pressing, Microwave and Ultrasound Treatments. *Biomass Conv. Bioref.* **2025**, *15*, 6483–6494. [CrossRef]
75. Piasecka, I.; Brzezińska, R.; Kalisz, S.; Wiktor, A.; Górska, A. Recovery of Antioxidants and Oils from Blackcurrant and Redcurrant Wastes by Ultrasound-Assisted Extraction. *Food Biosci.* **2024**, *57*, 103511. [CrossRef]
76. Ni, Q.; Gao, Q.; Yu, W.; Liu, X.; Xu, G.; Zhang, Y. Supercritical Carbon Dioxide Extraction of Oils from Two *Torreya Grandis* Varieties Seeds and Their Physicochemical and Antioxidant Properties. *LWT* **2015**, *60*, 1226–1234. [CrossRef]
77. *PN-EN ISO 5509:2001*; Vegetable and Animal Oils and Fats. Preparation of Fatty Acid Methyl Esters. Polish Committee for Standardization: Warsaw, Poland, 2001.
78. AOCS Official Method Cd 3d-63. Acid Value. In *Sampling and Analysis of Commercial Fats and Oils*; The American Oils Chemist's Society: Urbana, IL, USA, 2009.
79. AOCS Official Method Cd 8-53. Peroxide Value Acetic Acid (Chloroform Method). In *Sampling and Analysis of Commercial Fats and Oils*; The American Oils Chemist's Society: Urbana, IL, USA, 2003.
80. AOCS Official Method Cd 18-90. p-Anisidine Value. In *Official Methods and Recommended Practices of the AOCS.*; American Oil Chemists Society Press: Champaign, IL, USA, 2011.
81. Gao, X.; Ohlander, M.; Jeppsson, N.; Björk, L.; Trajkovski, V. Changes in Antioxidant Effects and Their Relationship to Phytonutrients in Fruits of Sea Buckthorn (*Hippophae rhamnoides* L.) during Maturation. *J. Agric. Food Chem.* **2000**, *48*, 1485–1490. [CrossRef] [PubMed]
82. Re, R.; Pellegrini, N.; Proteggente, A.; Pannala, A.; Yang, M.; Rice-Evans, C. Antioxidant Activity Applying an Improved ABTS Radical Cation Decolorization Assay. *Free Radic. Biol. Med.* **1999**, *26*, 1231–1237. [CrossRef] [PubMed]
83. Brand-Williams, W.; Cuvelier, M.E.; Berset, C. Use of a Free Radical Method to Evaluate Antioxidant Activity. *LWT* **1995**, *28*, 25–30. [CrossRef]

molecules

MDPI

Article

Optimization of Ultrasonic-Assisted Extraction of Diene Urushiol from Lacquer Tree Leaves Using Response Surface Methodology

Fengming Xia [1,†], Haojiang He [1,†], Jize Ma [1], Yutian Jin [1], Qing Qiao [1], Peng Long [1], Ping Li [2] and Rui Sun [1,*]

[1] College of Biological Science and Food Engineering, Southwest Forestry University, Kunming 650224, China; 16687002536@163.com (F.X.); hehj01@swfu.edu.cn (H.H.); Xanny73@163.com (J.M.); jinyutian2025@163.com (Y.J.); 15291762523@163.com (Q.Q.); lp18313786420@163.com (P.L.)

[2] Key Laboratory for Forest Resources Conservation and Utilization in the Southwest Mountains of China, Ministry of Education, Southwest Forestry University, Kunming 650224, China; liping2020@swfu.edu.cn

* Correspondence: sunrui@swfu.edu.cn

† These authors contributed equally to this work.

Abstract: Lacquer trees are an important economic tree species in China, and raw lacquer is its main secondary metabolite. Polyphenolic compounds are the primary components of raw lacquer, among which diene urushiol exhibits high inhibitory activity against the reverse transcriptase of acquired immunodeficiency syndrome (AIDS). Therefore, this study established and optimized the ultrasound-assisted extraction process of diene urushiol from lacquer tree leaves. Based on single-factor experiments on the number of extractions, extraction time, extraction temperature, and solvent to solid ratio, the Box–Behnken Design response surface methodology was employed to obtain the optimal extraction process, which included three extractions, an extraction time of 55 min, an extraction temperature of 50 °C, and a solvent to solid ratio of 10:1 mL/g. Under these conditions, the content of diene urushiol was 4.56 mg/g (FW), which bore no significant difference from the theoretical value of 4.69 mg/g (FW), indicating a good model fit. Therefore, response surface methodology (RSM) can be used to optimize the extraction process of diene urushiol from lacquer leaves. This method lays a solid foundation for the comprehensive development and utilization of lacquer tree resources.

Keywords: lacquer tree; diene urushiol; ultrasound-assisted extraction; response surface methodology (RSM)

1. Introduction

Lacquer trees, *Toxicodendron vernicifluum* (Stokes) F.A. Barkley, are deciduous trees belonging to the *Anacardiaceae* family and the *Toxicodendron* genus and are important, special economic trees in China [1]. Raw lacquer is the main secondary metabolite of lacquer trees and is regarded as a natural resin coating, often referred to as the "king of coatings". Urushiol is the primary component (comprising 40% to 70%) in raw lacquer and mainly includes saturated monoene, diene, and triene urushiols with C15 or C17 [2]. Urushiol possesses antioxidant [3], antitumor [4] antibacterial [5], and immune-regulating [6] properties. Studies have reported that diene urushiol exhibits high inhibitory activity against the reverse transcriptase of acquired immunodeficiency syndrome (AIDS), which can help prevent and treat AIDS [7].

Surveys indicate that over half of the individuals exposed to lacquer trees develop contact dermatitis [8]. Due to this particularity, the development and utilization of lacquer trees are limited, leading to their nickname, "toxic lacquer tree". Currently, the primary use of lacquer trees involves harvesting raw lacquer from the bark for coatings and extracting oil from their seeds [9]. Research indicates that lacquer tree leaves contain urushiol similar to those found in raw lacquer [2]. Consequently, lacquer tree leaves, being a renewable resource, can serve as raw materials for extracting urushiol for efficient utilization of lacquer tree plant resources.

Currently, the extraction methods mainly include solvent extraction via thermal reflux [10,11], ultrasound-assisted extraction [12], microwave-assisted extraction [13], and aqueous two-phase extraction [14]. However, thermal reflux has a long extraction time, microwave-assisted extraction has high energy consumption, and aqueous two-phase extraction consumes solvent because of complex phase separation [15]. Ultrasound-assisted extraction is more commonly used for extracting phenolic compounds because it employs the combined effects of cavitation, vibration, crushing, and stirring generated by ultrasound waves to disrupt cell walls, achieving efficient extraction of cellular contents. Therefore, ultrasound-assisted extraction can effectively solve the problems of long extraction time, high energy consumption, and low yield. However, there have been no reports on the ultrasound-assisted extraction of urushiol from lacquer tree leaves. When bioactive components are extracted from plants by ultrasonic-assisted extraction, the factors that can influence the extraction include extraction temperature, extraction time, the number of extractions, the solvent to solid ratio, etc. [16–18]. To attain optimum extraction efficiency, we conducted research on these parameters.

Response surface methodology (RSM) is a method that involves experimental design, model establishment, and data analysis to evaluate the interactions between factors across a region and to determine optimal conditions and response values [18,19]. It is commonly used in the extraction studies of plant components [20]. Therefore, this study used the renewable part of lacquer tree leaves as raw materials and optimized the ultrasound-assisted extraction process of diene urushiol using the RSM method, based on single-factor experiments (Figure 1). The purpose of this study is to improve the extraction efficiency of diene urushiol and make full use of lacquer tree plant resources. Ultimately, it lays the foundation for the comprehensive utilization of lacquer trees.

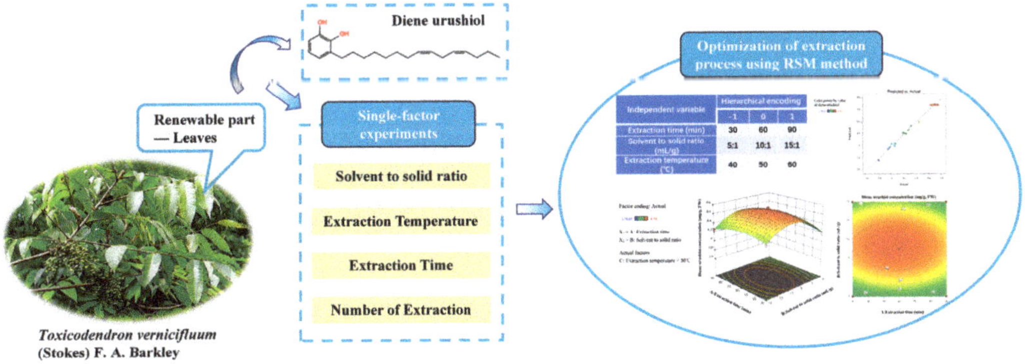

Figure 1. The experimental scheme of this study.

2. Results and Discussion

2.1. High-Performance Liquid Chromatography (HPLC) Analysis

HPLC analysis was conducted using the concentration of the diene urushiol standard as the x-axis (X, mg/mL) and the peak area as the y-axis (Y), resulting in a linear regression equation for the standard curve of diene urushiol:

$$y = 3590x - 27.327 \ (R^2 = 0.993) \quad (1)$$

A good linear relationship was observed within the range of 0.1847 to 0.9235 mg/mL (Figure 2).

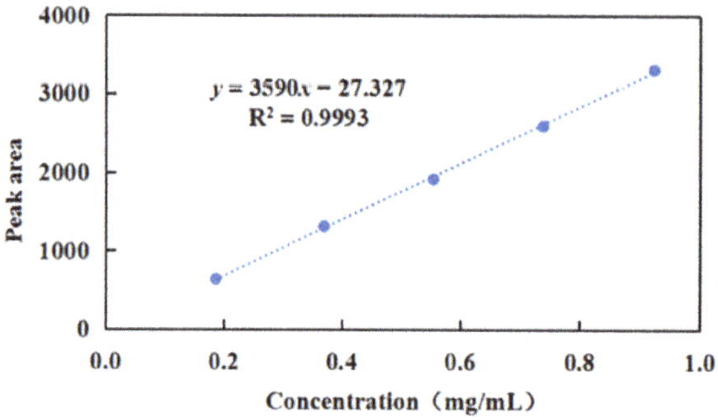

Figure 2. Standard curve of diene urushiol.

2.2. Results of Single-Factor Experiments

2.2.1. Effect of Solvent to Solid Ratio on the Extraction Content of Diene Urushiol

Increasing the solvent to solid ratio can enhance the solubility of the target components in the solvent, and the amount of extraction solvent significantly affects the extraction of the target components [21–23]. Sofi et al. reported optimal solvent to solid ratio of 8:1 mL/g for total phenolic compounds recovery from olive pomace oil [24]. A minimum solvent to solid ratio of 3:1 mL/g was found to ensure complete immersion of the sample powder, whereas a maximum ratio of 15:1 mL/g, which is approximately double that reported by Sofi et al., was sufficient for the extraction of diene urushiol. Therefore, in this study, the effects of four different solvent to solid ratios of 3:1, 5:1, 10:1 and 15:1 mL/g. The effect of the solvent to solid ratio on the content of diene urushiol was investigated, and the results are shown in Figure 3A. The results indicate that as the solvent to solid ratio increases from 3:1 to 10:1 mL/g, the content of diene urushiol significantly increases; however, when the ratio is further increased from 10:1 to 15:1 mL/g, the content of diene urushiol no longer shows significant changes. This suggests that at a solvent to solid ratio of 10:1 mL/g, diene urushiol is effectively extracted in the solvent. Chen et al. investigated ultrasonic-assisted extraction of flavonoids, polysaccharides, and eleutherosides from Acanthopanax senticosu, and the optimal solvent to solid ratio was 10:1 mL/g [25]. This result is similar to our study. Therefore, the solvent to solid ratio range of 5:1 to 15:1 mL/g was selected in the further RSM studies.

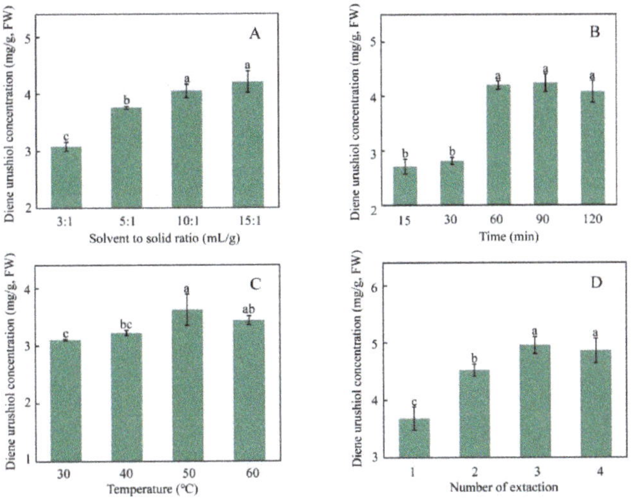

Figure 3. Effects of solvent to solid ratio (**A**), extraction time (**B**), extraction temperature (**C**), and number of extractions (**D**) on the content of diene urushiol. Different letters indicate the statistical significance at $p < 0.05$.

2.2.2. Effect of Extraction Time on the Extraction Content of Diene Urushiol

Increasing extraction time is beneficial for the extraction of effective components. However, it also increases the cost of time and energy [26]. Rosa et al. identified an optimal extraction time of 113.6 min within a 20–120 min range for total phenolic compounds extraction from grape peels [27]. On this basis, the present study investigated the effects of five different extraction times on the content of diene urushiol was investigated, and the results are shown in Figure 3B. From 15 to 30 min, the content of diene urushiol did not change with increasing time; however, from 30 to 60 min, there was a significant increase in the content of diene urushiol. After 60 min, the content no longer changed. This indicates that 60 min is sufficient for the complete extraction of diene urushiol. Therefore, the extraction time range of 30 to 90 min was selected in the following study.

2.2.3. Effect of Extraction Temperature on the Extraction Content of Diene Urushiol

Temperature has a significant impact on extraction performance, and an increase in ultrasonic temperature may lead to a higher diffusion coefficient of the target compounds and enhance their solubility [28]. Liao et al. determined an optimal temperature of 44.5 °C for total phenolic compounds extraction from purple eggplant peels [29]. Based on this, the present study investigated the effects of four different extraction temperatures of 30, 40, 50 and 60 °C on the content of diene urushiol, and the results are shown in Figure 3C. The content of diene urushiol significantly increased with the rise in extraction temperature from 30 to 50 °C. However, when the extraction temperature reached 50 °C, further increasing the temperature resulted in a decrease in the content of diene urushiol, possibly due to degradation of the compound at higher temperatures [30]. Diene urushiol degrades above 50 °C, which is similar to the thermal stability trends of curcumin and anthocyanins [29]. Therefore, it is believed that 50 °C is a suitable temperature for extracting diene urushiol. The extraction temperature range of 40 to 60 °C was selected in the following study. Although diene urushiol exhibits good bioactivity, its stability under long-term storage conditions still needs further study [31].

2.2.4. Effect of the Number of Extractions on the Extraction Content of Diene Urushiol

Before the active components are extracted completely, the content of active components can increase by increasing the number of extractions. However, as the number of extractions increases, more solvent is used, and more time is consumed [32]. Therefore, selecting the appropriate number of extractions is crucial. Wang et al. investigated the effect of the number of extractions on phenolic compound yields from Inula helenium within the number range of 1–4 [33]. Based on this, in the present study the effects of four different numbers of extractions at 1, 2, 3, and 4 on the diene urushiol content were studied, and the results are shown in Figure 3D. The content of diene urushiol significantly increased from extractions one to three. After the third extraction, the content of diene urushiol no longer changed, indicating that it was fully extracted at this point. Therefore, three extractions is energy-efficient and preferable.

2.3. *Establishment of the Response Surface Box–Behnken Design*

RSM design typically focuses on studying the interactions between continuous variable factors, while the number of extractions is often a discrete integer variable. Its variation lacks continuity, which may make it less suitable for fitting response surface models. The factor of extraction frequency was not included in the subsequent response surface optimization research. The effect of extraction frequency was only studied in single-factor experiments. Therefore, solvent to solid ratio, extraction time, and extraction temperature were selected for further optimization of the response surface model. In the Box–Behnken design (BBD) method of response surface methodology, the true values and coded levels of the independent variables are shown in Table 1. Seventeen experimental groups were obtained through response surface experimental design, and the results are shown in Table 2.

Table 1. Independent variables of BBD experiment and their levels.

Independent Variable	Symbol	Hierarchical Encoding		
		−1	0	1
Extraction time (min)	A	30	60	90
Solvent to solid ratio (mL/g)	B	5:1	10:1	15:1
Extraction temperature (°C)	C	40	50	60

Table 2. Experimental design and results of response surface analysis.

Serial Number	Extraction Time (min)	Solvent to Solid Ratio (mL/g)	Extraction Temperature (°C)	Diene Urushiol Content [1] (mg/g, FW)
1	90	10:1	60	4.05
2	60	10:1	50	4.65
3	60	10:1	50	4.71
4	60	10:1	50	4.74
5	30	10:1	40	3.94
6	90	5:1	50	4.23
7	60	15:1	60	3.96
8	90	15:1	50	4.27
9	30	15:1	50	4.31
10	60	15:1	40	4.04
11	60	10:1	50	4.62
12	90	10:1	40	4.18
13	30	5:1	50	4.21

Table 2. Cont.

Serial Number	Extraction Time (min)	Solvent to Solid Ratio (mL/g)	Extraction Temperature (°C)	Diene Urushiol Content [1] (mg/g, FW)
14	60	5:1	40	3.78
15	30	10:1	60	4.44
16	60	10:1	50	4.69
17	60	5:1	60	3.99

[1] Each value is the mean of triplicate measure, n = 3.

2.4. Model Fitting Statistics and Variance Analysis

Design Expert 11.0 software was used for analysis, and the quadratic multiple regression equation for the extraction content of diene urushiol (Y) is

$$Y = 4.68 - 0.0199A + 0.0456B + 0.0626C - 0.016AB - 0.1566AC - 0.0722BC - 0.1101A^2 - 0.3166B^2 + 0.4192C^2 \ (R^2 = 0.9876) \quad (2)$$

Among them, Y represents the extraction concentration of the diene urushiol, A represents extraction time (min), B represents solvent to solid ratio (mL/g), and C represents extraction temperature (°C).

The analysis of variance (ANOVA) was conducted for each factor related to the content of diene urushiol. In Table 3, it can be seen that the model is highly significant ($p < 0.0001$, f = 62.17), and the lack of fit is not significant ($p > 0.05$, f = 1.44). It can be observed that the linear terms of the solvent to solid ratio and extraction temperature have a significant effect on the diene urushiol content ($p < 0.05$). The two-way interaction between extraction time and extraction temperature ($p < 0.01$) has a greater impact on the diene urushiol content than that between solvent to solid ratio and extraction temperature ($p < 0.05$). When the interaction of extraction time and temperature is significant ($p < 0.01$), prolonging the extraction time at a high temperature will exacerbate the degradation phenomenon [34], which is similar to the result found in this study. The quadratic terms of the solvent to solid ratio and extraction temperature have an extremely significant effect on the diene urushiol content ($p < 0.0001$), and the quadratic term of the extraction time has a highly significant impact on the diene urushiol content ($p < 0.01$). A higher statistical correlation coefficient ($R^2 > 0.90$) indicates a good fit between experimental values and those obtained by the models [35]. The statistical correlation coefficient R^2 is 0.9876 in this study, which is an ideal observation ($R^2 > 0.90$) for the fitted model, proving that the experimental values fit well with the model values. The distribution of actual values approximates a straight line, further confirming the model's adequacy for predicting and analyzing experimental results in Figure 4.

Table 3. Analysis of variance of regression model.

Source	Sum of Squares	df	Mean Square	F-Value	p-Value
Model	1.4900	9	0.1658	62.1700	<0.0001 ***
A-Extraction time	0.0032	1	0.0032	1.1900	0.3110
B-Solvent to solid ratio	0.0166	1	0.0166	6.2400	0.0411 *
C-Extraction temperature	0.0314	1	0.0314	11.7700	0.0110 *
AB	0.0010	1	0.0010	0.3840	0.5551
AC	0.0981	1	0.0981	36.8000	0.0005 **
BC	0.0209	1	0.0209	7.8300	0.0266 *

Table 3. *Cont.*

Source	Sum of Squares	df	Mean Square	F-Value	p-Value
A^2	0.0511	1	0.0511	19.1500	0.0033 **
B^2	0.4220	1	0.4220	158.2700	<0.0001 ***
C^2	0.7400	1	0.7400	2.7751	<0.0001 ***
Residual	0.0187	7	0.0027		
Lack of fit	0.0097	3	0.0032	1.4400	0.3554
Pure error	0.0090	4	0.0022		
Cor total	1.5100	16			

* means significant at $p \leq 0.05$; ** means highly significant at $p \leq 0.01$; *** means extremely significant at $p \leq 0.0001$.

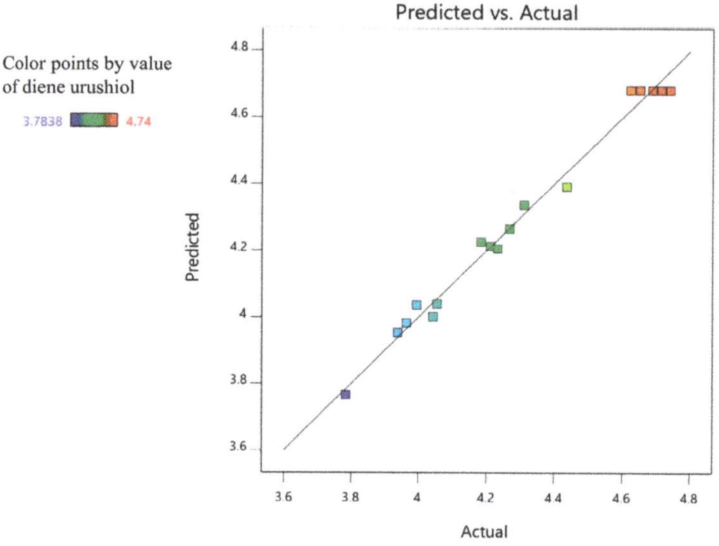

Figure 4. Predicted value and actual value of the contents for diene urushiol.

2.5. Analysis of Response Surface Interactions

The three-dimensional response surface plot can visually indicate the impact of each variable at different experimental levels and the type of interaction between two variables to determine the optimal conditions for maximizing the response value [36]. Steep curves in the three-dimensional response surface plot indicate significant interactions between the factors, while flat curves suggest that the effects of these factors are not significant [18,19,35]. In Figure 5A, it can be observed that when the solvent to solid ratio is constant, the 3D surface for the extraction time is not steep, indicating that the interaction between solvent to solid ratio and extraction time is not significant. Figure 5B,C show a peak in the 3D surface, with isopleths appearing elliptical. This indicates that the interaction between extraction time and extraction temperature, as well as the interaction between solvent to solid ratio and extraction temperature, has a significant impact on the content of diene urushiol. In Figure 5B, it can be observed that when the temperature varies from 40 to 60 °C, the content of diene urushiol increases and then decreases. This may be due to the increase in molecular diffusion rate with the rising temperature at lower temperatures, but further temperature increases can negatively affect the bioactive substances [37], leading to a decrease in the extraction content of diene urushiol. Additionally, it can be seen from Figure 5C that when the extraction temperature and/or the solvent to solid ratio are low or high, the extraction content of diene urushiol is low. A higher extraction amount of diene urushiol can be obtained only with moderate extraction temperature and solvent to solid ratio. At the same time, it can be observed that the surface steepness on the extraction temperature side is greater than that on the solvent to solid ratio side. This indicates that under the influence of their interaction, the extraction temperature has a greater impact on the extraction content of diene urushiol than the solvent to solid ratio. The results of the above analysis are consistent with the results of the variance analysis.

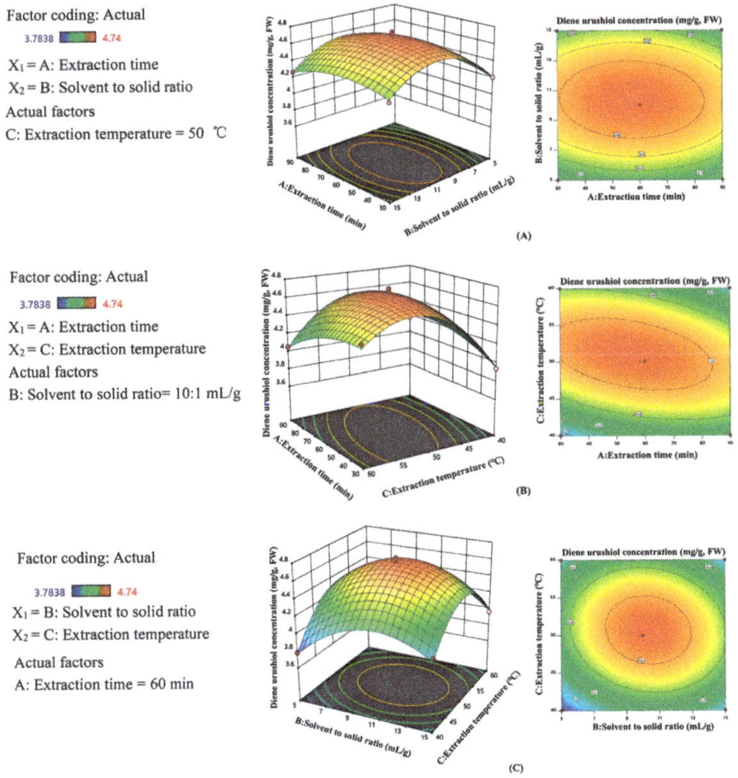

Figure 5. Interaction of different variables in the response surface. (**A**) extraction time and solvent to solid ratio, (**B**) extraction time and extraction temperature, (**C**) solvent to solid ratio and extraction temperature.

2.6. Model Validation

Through Design Expert 11.0 software, the optimal conditions predicted by the model were obtained as follows: extraction time of 55 min, extraction temperature of 51 °C, and solvent to solid ratio of 10.3:1 mL/g. Under these conditions, the predicted content of diene urushiol was 4.69 mg/g, FW. Considering the feasibility of practical operations, the extraction process for diene urushiol was adjusted to an extraction time of 55 min, extraction temperature of 50 °C, and solvent to solid ratio of 10:1 mL/g. Under this extraction process, the content of diene urushiol obtained was 4.56 mg/g, FW, which bore no significant difference compared to the model-predicted value of 4.69 mg/g, FW (RSD of 2.26%). The chromatogram of the sample solution obtained under optimal extraction conditions is presented in Figure 6.

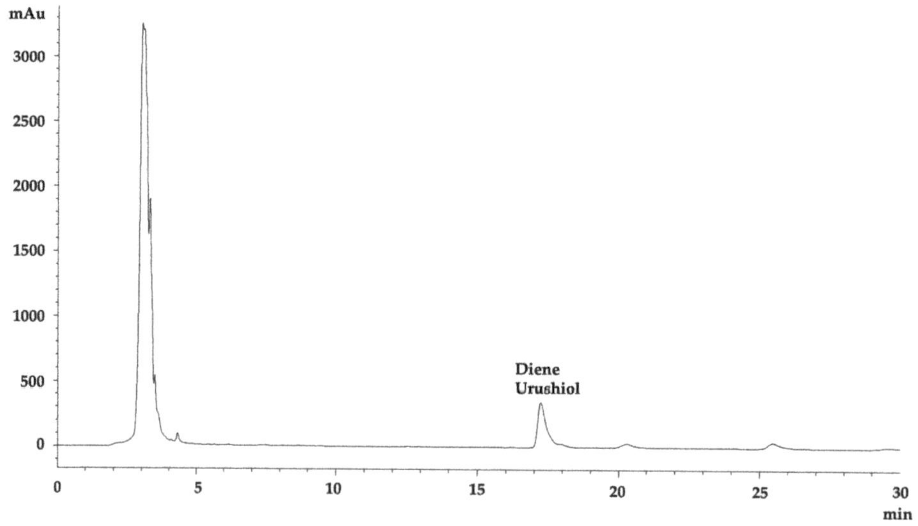

Figure 6. The chromatogram of the sample solution.

3. Materials and Methods

3.1. Materials and Reagents

The lacquer tree leaves were collected from wild lacquer trees in Dachongzi Village in August 2022, Wenshan City and Miao Autonomous Prefecture, Yunnan Province of China. After harvesting, they were immediately stored in a −80 °C freezer and ground into powder using liquid nitrogen during the experiment. The standard of diene urushiol with 98% purity was purchased from Shanghai Dande Biotechnology Co., Ltd. (Shanghai, China). Analytical-grade ethanol was purchased from Guangdong Guanghua Technology Co., Ltd. (Shantou, China). Chromatographic acetonitrile was purchased from Beijing Meiliade Technology Co., Ltd. (Beijing, China), and ultrapure water was made by a water purifier (PP010XXM1, ELGA, Warwickshire, UK) for all experiments.

3.2. Quantitative Determination of Diene Urushiol by HPLC

The Agilent high-performance liquid chromatography system (Agilent Technologies, Santa Clara, CA, USA) was used for the determination of diene urushiol, comprising an automatic injector, a quaternary pump, a DAD UV detector, and an EC-C18 column (250 mm × 4.6 mm; 4 μm). The mobile phase consisted of acetonitrile and water (83:17), with a flow rate of 0.7 mL/min, and the analytical time for a single sample was 30 min. The column temperature was maintained at 40 °C, the detection wavelength was set at 225 nm, and the injection volume was 2 μL. A precise amount of diene urushiol standard was accurately weighed and gradient-diluted using chromatographic methanol (Beijing Bailingwei Technology Co., Ltd., Beijing, China) to obtain the standard solution of varying concentrations (0.1847, 0.3694, 0.5541, 0.7388 and 0.9235 mg/mL). According to the HPLC procedure, the peak areas corresponding to different concentrations of the standard solutions were detected, and a standard curve was established to show the linear relationship between peak area and diene urushiol content.

3.3. Single-Factor Experimental Design

100% ethanol was used as the extraction solvent, and ultrasonic-assisted extraction was employed to extract diene urushiol from lacquer tree leaves under specific conditions of solvent to solid ratio, extraction time, extraction temperature, and extraction frequency. Fresh lacquer tree leaves were ground into powder using liquid nitrogen, and 1.00 g was weighed and placed in a centrifuge tube. Corresponding volumes of anhydrous ethanol solution were added according to specific solvent to solid ratios (3:1, 5:1, 10:1, and 15:1 mL/g). The mixture was subjected to ultrasonic extraction at specific temperatures (30, 40, 50, and 60 °C) in an ultrasonic bath (SK5210HP, Shanghai Kudos Ultrasonic Instrument Co., Ltd., Shanghai, China) for specified durations (15, 30, 60, 90, and 120 min). It was then centrifuged at 12,000 rpm for 10 min in a centrifuge (R134a, Eppendorf AG, Hamburg, Germany), and the supernatant was collected for use. The extraction was conducted with a certain frequency (1, 2, 3, and 4 times). All the supernatants were mixed, then concentrated to dryness by using rotary evaporation. Then, the residue was dissolved in 2 mL of anhydrous ethanol solution for HPLC analysis. All sample solutions were filtered through a membrane with a pore size of 0.22 μm before conducting HPLC analysis. Each treatment was repeated three times.

3.4. *Response Surface Design*

The Box–Behnken design (BBD), an important experimental design method in RSM, is widely used in scientific research and industrial production optimization due to its efficiency and accuracy [31]. Based on single-factor experiments, extraction time (A), solvent to solid ratio (B), and extraction temperature (C) were identified as three significant factors affecting the content of diene urushiol. Therefore, a response surface methodology (RSM-BBD) experimental design was conducted using these three factors at three levels. The extraction content of diene urushiol was set as the response value, and a second-order polynomial model was established to determine the regression coefficients. The experiments conducted based on the RSM experimental design, the diene urushiol contents were determined, and variance analysis was conducted to test the significance of the model concerning the dependent variable ($p < 0.05$). Three-dimensional response surface curves were also plotted to analyze the interaction effects of the factors on the response value. Finally, the optimal ultrasonic extraction process parameters were determined and validated by the response surface prediction model.

3.5. *Statistical Analysis*

Statistical analysis of experimental data was performed using Excel software (2019, Microsoft Office, Redmond, WA, USA), followed by Duncan's multiple range test and one-way ANOVA using IBM SPSS Statistics 26.0 (Statistical Product Service Solutions, Armonk, NY, USA). Experimental design, ANOVA, modelling, and predictive optimization were carried out using Design Expert 11.0 software (Stat Ease, Inc., Minneapolis, MN, USA).

4. Conclusions

This study used lacquer tree leaves as experimental materials and optimized the effects of three extraction factors (solvent to solid ratio, extraction time, and extraction temperature) on the ultrasonic-assisted extraction of diene urushiol using the Box–Behnken Design method of response surface methodology. The results indicated that the optimal extraction conditions were as follows: extraction time of 55 min, extraction temperature of 50 °C, and solvent to solid ratio of 10:1 mL/g. Under these extraction conditions, the content of diene urushiol was 4.56 mg/g (FW), while the theoretical value was 4.69 mg/g (FW). They showed no significant difference, with an RSD of 2.82% and a good model fit. Therefore, this method can be used to optimize the ultrasonic-assisted extraction process of diene urushiol from lacquer tree leaves, providing an effective approach for the comprehensive and sustainable development and utilization of lacquer tree resources. In the future, scale-up and purified experiments can be carried out to obtain pure diene urushiol for pharmaceutical applications.

Author Contributions: Conceptualization, R.S.; methodology, F.X., H.H. and R.S.; validation, Y.J. and J.M.; formal analysis, Q.Q., Y.J., P.L. (Peng Long) and H.H.; investigation, F.X., H.H. and J.M.; resources, R.S.; data curation, Q.Q., P.L. (Peng Long) and J.M.; writing—original draft preparation, F.X. and H.H.; writing—review and editing, R.S.; visualization, F.X., H.H., P.L. (Ping Li) and R.S.; supervision, P.L. (Ping Li) and R.S.; project administration, P.L. (Ping Li) and R.S. All authors have read and agreed to the published version of the manuscript.

Funding: The author(s) declare that financial support was received for the research, authorship, and/or publication of this article. This study was supported by Yunnan Major Science and Technology Projects (202302AE090002-03), Yunnan Xingdian Talent Support Program (YUWR-CYJS-2020-018), the Project of Yunnan Provincial Department of Education Science Research Fund (2022J0503), the Opening Project of Key Laboratory for Forest Resources Conservation and Utilization in the Southwest Mountains of China, Ministry of Education (KLESWFU201903, KLESWFU-202010), Yunnan Fundamental Research Projects (202201AU070072 and 202301AT070216), Foundation of Yunnan Agricultural Basic Research (202101BD070001-033), and the Scientific Research Fund Project of Southwest Forestry University.

Institutional Review Board Statement: Not applicable.

Informed Consent Statement: Not applicable.

Data Availability Statement: The data presented in this study are available in the article.

Acknowledgments: The authors express their gratitude to the Key Laboratory for Forest Resources Conservation and Utilization in the Southwest Mountains of China, Ministry of Education, Southwest Forestry University, for furnishing the laboratory facilities to carry out the research. All authors recognize the suggestions of the reviewers for enhancing the manuscript.

Conflicts of Interest: The authors declare no conflicts of interest.

References

1. Li, M.C.; Zhang, Y.Q.; Meng, C.W.; Gao, J.G.; Xie, C.J.; Liu, J.Y.; Xu, Y.N. Traditional uses, phytochemistry, and pharmacology of *Toxicodendron vernicifluum* (Stokes) F.A. Barkley—A review. *J. Ethnopharmacol.* **2021**, *276*, 113476. [CrossRef]
2. Du, Y.; Oshima, R.; Kumanotani, J. Reversed-phase liquid chromatographic separation and identification of constituents of uroshiol in the sap of the lac tree, rhus vernicifera. *J. Chromatogr. A* **1984**, *284*, 463–473. [CrossRef]
3. Lim, K.T.; Hu, C.; Kitts, D.D. Antioxidant activity of a *Rhus verniciflua* stokes ethanol extract. *Food Chem. Toxicol.* **2001**, *39*, 229–237. [CrossRef] [PubMed]
4. Jeong, J.H.; Ryu, J.H. Urushiol V suppresses cell proliferation and enhances antitumor activity of 5-fu in human colon cancer cells by downregulating foxm1. *Biomol. Ther.* **2022**, *30*, 257–264. [CrossRef]
5. Kim, J.S.; Kwon, Y.S.; Chun, W.J.; Kim, T.Y.; Sun, J.; Yu, C.Y.; Kim, M.J. *Rhus verniciflua* stokes flavonoid extracts have anti-oxidant, anti-microbial and α-glucosidase inhibitory effect. *Food Chem.* **2010**, *120*, 539–543. [CrossRef]
6. Kang, S.Y.; Kang, J.Y.; Oh, M.J. Antiviral activities of flavonoids isolated from the bark of *Rhus verniciflua* stokes against fish pathogenic viruses in vitro. *J. Microbiol.* **2012**, *50*, 293–300. [CrossRef]
7. Kim, M.J.; Choi, W.C.; Barshinikov, A.M.; Kobayashi, A. Inhibitory effects of the extract of *Rhus verniciflua* stokes on the reverse transcriptase of aids. *Korean J. Med. Crop Sci.* **2002**, *10*, 284–287.
8. Resnick, S.D. Poison-ivy and poison-oak dermatitis. *Clin. Dermatol.* **1986**, *4*, 208–212. [CrossRef]
9. Reda, F.M.; Alagawany, M.; Alsolami, A.M.; Mahmoud, H.K.; Salah, A.S.; Momenah, M.A.; Alshahrani, M.Y.; EL-Tarabily, K.A.; Saleh, R.M. The incorporation of sumac seed powder (*Rhus coriaria* L.) into the diet of quail breeders as a novel feed additive. *Poult. Sci.* **2024**, *103*, 103593. [CrossRef] [PubMed]
10. Park, K.Y.; Jung, G.O.; Lee, K.T.; Choi, J.; Choi, M.Y.; Kim, G.T.; Jung, H.J.; Park, H.J. Antimutagenic activity of flavonoids from the heartwood of *Rhus verniciflua*. *J. Ethnopharmacol.* **2004**, *90*, 73–79. [CrossRef]

11. Lee, J.D.; Huh, J.E.; Jeon, G.S.; Yang, H.R.; Woo, H.S.; Choi, D.Y.; Park, D.S. Flavonol-rich rvhxr from *Rhus verniciflua* stokes and its major compound fisetin inhibits inflammation-related cytokines and angiogenic factor in rheumatoid arthritic fibroblast-like synovial cells and in vivo models. *Int. Immunopharmacol.* **2009**, *9*, 268–276. [CrossRef] [PubMed]
12. Garcia-Oliveira, P.; Chamorro, F.; Simal-Gandara, J.; Proetp, M.A.; Cassani, L. Improving phenolic compound extraction from *arnica montana* flowers through multivariate optimization of heat and ultrasound-assisted methods. *Sustain. Chem. Pharm.* **2024**, *41*, 101722. [CrossRef]
13. Mokaizh, A.A.B.; Nour, A.H.; Ali, G.A.; Ukaegbu, C.I.; Hawege, E.F. Eco-friendly and efficient extraction of phenolic compounds from Commiphora gileadensis bark using microwave-assisted extraction. *J. Ind. Eng. Chem.* **2009**, *142*, 321–328. [CrossRef]
14. Wang, Y.; Wang, J.L.; Chen, S.J.; Zhang, Y.; Wu, J. Study on Aqueous Two-Phase Extraction of Teapolypheols from in Green Tea. *Key Eng. Mater.* **2022**, *915*, 101–106. [CrossRef]
15. Xiaolong, J.; Qiang, P.; Yuepeng, Y.; Fang, L.; Min, W. Extraction and physicochemical properties of polysaccharides from *Ziziphus Jujuba cv. Muzao* by ultrasound-assisted aqueous two-phase extraction. *Int. J. Biol. Macromol.* **2018**, *108*, 541–549. [CrossRef]
16. Li, J.; Chen, Z.; Shi, H.; Yu, J.; Huang, G.; Huang, H. Ultrasound-assisted extraction and properties of polysaccharide from Ginkgo biloba leaves. *Ultrason. Sonochem.* **2023**, *93*, 106295. [CrossRef]
17. Shen, L.; Pang, S.; Zhong, M.; Sun, Y.; Qayum, A.; Liu, Y.; Rashid, A.; Xu, B.; Liang, Q.; Ma, H.; et al. A comprehensive review of ultrasonic assisted extraction (UAE) for bioactive componentts: Principles, advantages, equipment, and combined technologies. *Ultrason. Sonochem.* **2023**, *101*, 106646. [CrossRef] [PubMed]
18. Srivastava, S.; Pandey, V.K.; Tripathi, A.; Singh, R.; Dash, K. Ultrasound-assisted oxalate reduction vs. conventional methods: A comparative study in elephant foot yam (*Amorphophallus paeoniifolius*) and process optimization using response surface methodology. *Food Humanit.* **2024**, *2*, 100217. [CrossRef]
19. Sai-Ut, S.; Kingwascharapong, P.; Mazumder, M.A.R.; Rawdkuen, S. Optimization of extraction of phenolic compounds and antioxidants from passion fruit and rambutan seeds using response surface methodology. *J. Agric. Food Res.* **2023**, *14*, 100888. [CrossRef]
20. Heleno, S.A.; Diz, P.; Prieto, M.A.; Barros, L.; Rodrigues, A.; Barreiro, M.F.; Ferreira, I.C.F.R. Optimization of ultraso-und-assisted extraction to obtain mycosterols from *Agaricus bisporus* L. by response surface methodology and comparison with conventional Soxhlet extraction. *Food Chem.* **2016**, *197*, 1054–1063. [CrossRef]
21. Xu, Y.; Pan, S. Effects of various factors of ultrasonic treatment on the extraction yield of all-trans-lycopene from red grapefruit (*Citrus paradise Macf.*). *Ultrason. Sonochem.* **2013**, *20*, 1026–1032. [CrossRef]
22. Zhang, D.Y.; Wan, Y.; Xu, J.Y.; Wu, G.H.; Li, L.; Yao, X.H. Ultrasound extraction of polysaccharides from mulberry leaves and their effect on enhancing antioxidant activity. *Carbohydr. Polym.* **2016**, *137*, 473–479. [CrossRef]
23. Xu, D.P.; Zheng, J.; Zhou, Y.; Li, Y.; Li, S.; Li, H.B. Ultrasound-assisted extraction of natural antioxidants from the flower of *Limonium sinuatum*: Optimization and comparison with conventional methods. *Food Chem.* **2017**, *217*, 552–559. [CrossRef]
24. Chen, X.; Jia, X.; Yang, S.; Zhang, G.; Li, A.; Du, P.; Liu, L.; Li, C. Optimization of ultrasonic-assisted extraction of flavonoids, polysaccharides, and eleutherosides from *Acanthopanax senticosus* using response surface methodology in development of health wine. *LWT* **2022**, *165*, 113725. [CrossRef]
25. Sofia, C.; Constantina, T. Optimization of ultrasound-assisted extraction of oil from olive pomace using response surface technology: Oil recovery, unsaponifiable matter, total phenol content and antioxidant activity. *LWT-Food Sci. Technol.* **2017**, *79*, 178–189. [CrossRef]
26. Liu, L.; Li, J.; Tian, J.; Zhou, Z.; Gao, J.; Qin, L.; Qiu, J. Evaluation of six phenolic compounds content and antifungal activity against wood-decay fungi of tobacco waste extracts obtained by ultrasound-assisted extraction. *Sustain. Chem. Pharm.* **2024**, *41*, 101718. [CrossRef]
27. Rosa, A.M.; Basile, T.; Alba, V.; Gasparro, M.; Giannandrea, M.A.; Debiase, G.; Genghi, R.; Antonacci, D. Optimized ultrasonic-assisted extraction of phenolic antioxidants from grape (*Vitis vinifera* L.) skin using response surface methodology. *J. Food Sci. Technol.* **2019**, *56*, 4417–4428. [CrossRef]
28. Li, Q.; Yu, N.; Wang, Y.; Sun, Y.; Lu, K.; Guan, W. Extraction optimization of *Bruguiera gymnorrhiza* polysaccharides with radical scavenging activities. *Carbohydr. Polym.* **2013**, *96*, 148–155. [CrossRef]
29. Liao, J.; Xue, H.; Li, J. Extraction of phenolics and anthocyanins from purple eggplant peels by multi-frequency ultrasound: Effects of different extraction factors and optimization using uniform design. *Ultrason. Sonochem.* **2022**, *90*, 1350–4177. [CrossRef]
30. Xie, J.H.; Shen, M.Y.; Xie, M.Y.; Nie, S.P.; Chen, Y.; Li, C.; Huang, D.F.; Wang, Y.X. Ultrasonic-assisted extraction, antimicrobial and antioxidant activities of cyclocarya paliurus (batal.) iljinskaja polysaccharides. *Carbohydr. Polym.* **2012**, *89*, 177–184. [CrossRef]
31. Han, G.G.D.; Li, H.; Grossman, J.C. Optically-controlled long-term storage and release of thermal energy in phase-change materials. *Nat. Commun.* **2017**, *8*, 1446. [CrossRef]
32. Chen, R.; Li, Y.; Dong, H.; Liu, Z.; Li, S.; Yang, S.; Li, X. Optimization of ultrasonic extraction process of polysaccharides from *Ornithogalum Caudatum Ait* and evaluation of its biological activities. *Ultrason. Sonochem.* **2012**, *19*, 1160–1168. [CrossRef] [PubMed]

33. Wang, J.; Zhao, Y.M.; Tian, Y.T.; Yan, C.L.; Guo, C.Y. Ultrasound-assisted extraction of total phenolic compounds from *Inula helenium*. *Sci. World J.* **2013**, *2013*, 157527. [CrossRef]
34. He, S.; Wang, X.; Chen, J.; Li, X.; Gu, W.; Zhang, F.; Cao, G.; Yu, J. Optimization of the ultrasonic-assisted extraction Technology of Steroidal Saponins from *Polygonatum kingianum* Collett & Hemsl and evaluating its quality planted in different areas. *Molecules* **2022**, *27*, 1463. [CrossRef]
35. Tirado-Kulieva, V.A.; Sánchez-Chero, M.; Yarlequé, M.V.; Aguilar, G.F.V.; Carrión-Barco, G.; Santa Cruz, A.G.Y. An overview on the use of response surface methodology to model and optimize extraction processes in the food industry. *Curr. Res. Nutr. Food Sci. J.* **2021**, *9*, 745–754. [CrossRef]
36. Lee, L.S.; Lee, N.; Kim, Y.H.; Lee, C.H.; Hong, S.P.; Jeon, Y.W.; Kim, Y.E. Optimization of ultrasonic extraction of phenolic antioxidants from green tea using response surface methodology. *Molecules* **2013**, *18*, 13530–13545. [CrossRef] [PubMed]
37. Senrayan, J.; Venkatachalam, S. Optimization of ultrasound-assisted solvent extraction (UASE) based on oil yield, antioxidant activity and evaluation of fattyacid composition and thermal stability of *Coriandrum sativum* L. Seed oil. *Food Sci. Biotechnol.* **2019**, *28*, 377–386. [CrossRef]

molecules

MDPI

Article

Optimization of Deep Eutectic Solvent-Based Ultrasound-Assisted Extraction of Bioactive Compounds from Maca Leaves Using the Taguchi Method

Eun Ji Lee and Kyung Young Yoon *

Department of Food and Nutrition, Yeungnam University, Gyeongsan 38541, Republic of Korea
* Correspondence: yoonky2441@ynu.ac.kr; Tel.: +82-53-810-2878

Abstract: This study was conducted to identify the optimal conditions and evaluate the feasibility of deep eutectic solvent (DES)-based ultrasound-assisted extraction (UAE) for utilizing maca (*Lepidium meyenii*) leaves, an agricultural by-product, as functional materials. The extraction parameters influencing the recovery of saponins and polyphenols, which are major bioactive compounds, were analyzed using the Taguchi method. Results: Signal-to-noise ratios and analysis of variance indicated that the liquid–solid ratio was the most critical factor for optimizing the extraction process. The optimal extraction conditions were determined to be a liquid–solid ratio of 40 mL/g, a water content in DES of 30%, an extraction time of 30 min, and an ultrasonic power of 300 W in the DES system consisting of choline chloride and glycerin in the molar ratio of 1:2. Maca leaf extract obtained under optimized DES-based UAE conditions exhibited higher bioactive compounds content and antioxidant activity compared with that obtained by hot water extraction. Therefore, the DES-based UAE method is a promising, eco-friendly alternative for extracting bioactive compounds from maca leaves.

Keywords: maca leaves; deep eutectic solvent; ultrasound-assisted extraction; saponin; polyphenol; Taguchi method

1. Introduction

Maca (*Lepidium meyenii*), also known as Peruvian ginseng, is a cruciferous crop that grows in the Andes region of South America. Maca is hardy, can grow in inhospitable natural environments, and is resistant to strong winds, cold, and drought [1]. Maca roots contain polyphenols, tannins, saponins, steroids, and a variety of organic compounds, including macaenes, macamides, glucosinolates, isothiocyanates, and alkaloids [2]. These components have been reported to act alone or synergistically to exhibit biological activities, which include neuro and liver protection, antioxidant, anti-fatigue, anti-cancer, and anti-osteoporosis properties, and immune regulation [3]. Accordingly, the consumption of maca root is continuously increasing; however, the maca leaves, a by-product generated during maca processing, are discarded as waste [4]. According to recent study on the bioactive compounds and biological activities of methanol extracts of maca leaves, they have a higher content of saponins, phenols, and flavonoids than maca root extracts [5]. In addition, the leaves contain various polyphenols, such as kuwanone-G, oxypaenoniflorin, laricitrin, and ligustroflavone [5,6]. Moreover, the aerial parts of maca plants (leaf, stem, and inflorescence) contain glucosinolates, macamides, and saponins at ranges of 31.4–36.2, 0.451–0.703, and 33.2–51.8, mg/g (dry weight), respectively, at different growth stages [7].

However, although maca leaves contain a variety of bioactive compounds, they are currently not viewed as a potential source of functional substances [7]. Therefore, research is needed to explore the potential of maca leaves discarded as waste as a functional material.

To recover bioactive substances from a plant matrix, an appropriate extraction process is required. Recently, eco-friendly extraction technologies, such as supercritical fluid extraction, ultrasound-assisted extraction (UAE), and microwave-assisted extraction (MAE) have been applied to replace conventional extraction methods [8]. Among these technologies, UAE is widely used, where ultrasound can improve the mass transfer of compounds from plant substrates to solvents by generating cavitation bubbles that disrupt the plant cell walls. This improves the extraction yield and saves time, energy, and solvents compared to conventional extraction methods [9]. In addition, interest in the use of eco-friendly solvents that do not negatively impact the environment and humans is increasing; among these, deep eutectic solvent (DES), which has low toxicity, is the most promising [10]. A DES can be simply prepared as an eutectic mixture containing a mixture of a hydrogen bond donor and acceptor molecules. That is, when two chemicals are stirred together at a specific temperature, they mutually induce melting and form hydrogen bonds, resulting in a transparent solution of DES at a temperature significantly lower than the melting points of the individual substances [11]. Utilizing DES as an extraction solvent facilitates the isolation of target components from plant materials through hydrogen bonding and electrostatic interactions between DES constituents and the target molecules [12]. DES offers several advantages, including simple production, low toxicity, environmental compatibility, high biocompatibility and biodegradability, and low cost. Furthermore, DES can be synthesized with diverse chemical properties by altering one or two of its components, making it a versatile solvent for the extraction of various compounds, thereby overcoming the limitations of existing extraction solvents [13]. In addition, DES can be efficiently used to extract various nonpolar and polar bioactive substances from natural plant resources by using a combination of thermal extraction, UAE, and MAE [14]. In particular, many studies have been conducted to extract various bioactive components such as polysaccharides and polyphenols from plants using DES-based UAE and to improve extraction efficiency [12,15,16]. However, no studies have been conducted to extract bioactive substances such as saponins and polyphenols from maca leaves.

Therefore, this study aimed to determine the optimal DES-based UAE conditions to improve the extraction efficiency of bioactive compounds, including polyphenols and saponins, from maca leaves. For this purpose, the extraction conditions, including the liquid–solid ratio, water content in DES, extraction time, and ultrasonic power, were optimized using Taguchi's methodology. In addition, the bioactive compounds and antioxidant activities of the maca leaf extracts obtained under the optimized DES-based UAE conditions were compared with those from hot water extraction (HWE), one of the representative conventional extraction methods.

2. Results and Discussion

2.1. Total Saponin and Polyphenol Contents

To determine the optimal conditions for DES-based UAE for the maximum recovery of bioactive compounds from maca leaves, four factors (liquid–solid ratio, water content in DES, extraction time, and ultrasonic power) were set to three levels, and sample extraction was performed according to the L_9 (3^4) orthogonal matrix. The results of orthogonal experiments are shown in Table 1. The total saponin content (TSC) of the extracts obtained under various conditions was 16.43–33.43 mg oleanolic acid equivalent (OAE)/g dry weight (DW); the highest content of extracted TSC was achieved using an liquid–solid ratio of 40 mL/g, 30% water content, 20 min extraction time, and 180 W ultrasonic power. The total polyphenol content (TPC) obtained according to each parameter was found to be 3.10–6.32 mg gallic acid equivalent (GAE)/g DW, and runs 7–9 showed higher TPC. For runs 7–9, the water content, extraction time, and ultrasonic power showed some differences, but the liquid–solid ratio was the same at 40 mL/g. In addition, the TSC and TPC increased with the increasing liquid–solid ratio, and the highest TSC and TPC were obtained at a liquid–solid ratio of 40 mL/g, which is consistent with the results reported by Liu et al. [17].

Table 1. L_9 orthogonal designs, levels of four factors, and experimental results.

Run No.	Factors				Responses		S/N Ratios	
	Liquid–Solid Ratio (mL/g)	Water Content (%)	Extraction Time (min)	Ultrasonic Power (W)	TSC (mg OAE/g DW)	TPC (mg GAE/g DW)	TSC	TPC
1	20	10	10	180	16.43 ± 0.28	3.12 ± 0.18	24.31	9.84
2	20	20	20	240	21.69 ± 1.27	3.10 ± 0.09	26.69	9.82
3	20	30	30	300	27.60 ± 0.11	4.43 ± 0.07	28.82	12.93
4	30	10	20	300	24.85 ± 0.60	4.74 ± 0.70	27.90	13.27
5	30	20	30	180	27.95 ±1.28	4.53 ± 0.27	28.91	13.08
6	30	30	10	240	30.26 ± 2.19	4.87 ± 0.74	29.56	13.50
7	40	10	30	240	27.84 ± 2.53	6.05 ± 0.17	28.80	15.63
8	40	20	10	300	31.98 ± 2.05	6.32 ± 0.32	30.05	15.99
9	40	30	20	180	33.43 ± 3.48	6.31 ± 0.24	30.37	15.99
*K*1 (TSC)	21.91	23.04	26.23	25.94				
*K*2 (TSC)	27.69	27.21	26.65	26.59				
*K*3 (TSC)	31.08	30.43	27.79	28.14				
Delta value (TSC)	9.17	7.39	1.56	2.20				
*K*1 (TPC)	3.55	4.64	4.77	4.65				
*K*2 (TPC)	4.71	4.65	4.72	4.67				
*K*3 (TPC)	6.23	5.21	5.00	5.16				
Delta value (TPC)	2.68	0.57	0.28	0.51				

TSC, total saponin content; OAE, oleanolic acid equivalent; TPC, total polyphenol content; GAE, gallic acid equivalent; DW, sample dry weight; and S/N, signal-to-noise. Experimental value is represented as mean ± standard deviation of triplicate. K_i^A = (Σ response at A_i)/3, where A is each factor-variable level. Delta value: Refers to the result of extreme analysis, $\Delta = \max\{K_i^A\} - \min\{K_i^A\}$.

The liquid–solid ratio is reported to substantially impact the solvent consumption, recovery of target compounds, and solvent regeneration costs. In general, as the liquid–solid ratio increases, the yield of phenolic compounds increases and the extraction time is shortened. This is because a larger volume of solvent increases the contact area with the target material, accelerating the diffusion of the solute into the solvent [18]. However, considering the solvent and target compound recovery, it is necessary to establish an optimal range that balances extraction yield and solvent consumption [19].

In general, the extraction ability of DES for bioactive substances is related to its physical and chemical properties, such as hydrogen bonding, viscosity, polarity, and pH [20]. In particular, adding water to DES, which is highly viscous, can lower its viscosity; however, this may increase its polarity and weaken the hydrogen bonds between acceptor and donor molecules [21,22]. Therefore, to promote the transfer of the target ingredient to the solvent, it is necessary to adjust the amount of water added, that is, the viscosity of DES [23]. The *K* value showed that the TSC and TPC increased as the water content of the DES increased. The highest TSC and TPC were observed at 30.43 mg OAE/g DW and 5.21 mg GAE/g DW, respectively, when the water content added to DES was 30%. Considering these results, the saponins and polyphenols extracted from the maca leaves using DES prepared with choline chloride and glycerin, which are polar substances, and the 30% water content in DES, are suggested to have increased the extraction yield by appropriately lowering the viscosity and increasing the polarity of the solvent.

2.2. Analysis of Main Effect on Bioactive Compounds Recovery

To examine the effect of extraction conditions on the recovery of the bioactive compounds in the maca leaf extract, the effect of each factor was calculated based on an orthogonal test (Table 1). The delta value represents the relative effect of each factor on the response as the difference between the maximum and minimum average response values for each factor. Based on the average delta value for the responses obtained at each factor-variable level in this study, the TSC was influenced by the parameters in the following order: liquid–solid ratio (9.17 mg OAE/g DW), water content (7.39 mg OAE/g DW), ultrasonic power (1.56 mg OAE/g DW), and extraction time (2.20 mg OAE/g DW). Similarly, the TPC was affected by the parameters in the following order: liquid–solid ratio (2.68 mg GAE/g DW), water content (0.57 mg GAE/g DW), ultrasonic power (0.28 mg GAE/g DW), and extraction time (0.51 mg GAE/g DW).

In the Taguchi method, the experimental conditions with the least variability are identified as the optimal conditions, and this variability is expressed as the signal-to-noise (S/N) ratio. Since variable characteristics are inversely proportional to the S/N ratio, the experimental conditions with the highest S/N ratios are considered optimal [24]. Additionally, depending on the objective of the experiment, the results can be interpreted using the most appropriate method between nominal-is-best, smaller-the-best, and larger-the-best characteristics [18]. As shown in Table 1, in this study, the TSC showed the highest S/N ratio at run 9 (30.37), and the TPC showed the highest S/N ratio at runs 8 and 9 (15.99).

In the main effect plot for the S/N ratios, the mean S/N ratios for each factor were plotted against the test levels for each factor parameter (Figure 1). The main effects plot of TSC indicates that the liquid–solid ratio had the largest influence, followed by the water content, while the extraction time and ultrasonic power had very slight impacts (Figure 1A). Similarly, the TPC was greatly influenced by the liquid–solid ratio, with the effects of the other parameters being very slight (Figure 1B).

2.3. Analysis of Variance (ANOVA) Results

The Taguchi experimental design uses orthogonal arrays and ANOVA as analysis tools. After conducting a minimal number of experiments using the orthogonal array method, the effect of the specific factors studied on the response is measured using ANOVA. The ANOVA was conducted at a 95% confidence level to identify the significant variables and quantify their effects on the responses (Table 2). The ANOVA results indicated that all the selected factors were substantial parameters of the DES-based UAE process at the 95% confidence level ($p < 0.05$). Considering the percentage contribution of each factor on extraction, the results revealed that the liquid–solid ratio had the most influence on the TSC and TPC among the process parameters, probably owing to its highest percentage contribution and low *p*-value ($p < 0.001$). The liquid–solid ratio exhibited the highest percentage contribution (51.85%; $p < 0.001$) on the extracted TSC, followed by the water content (33.06%; $p < 0.001$), ultrasonic power (3.08%; $p = 0.003$), and extraction time (1.58%; $p = 0.042$). In the case of the TPC, the percentage contribution in the extraction process was in the same order as for the TSC; however, the liquid–solid ratio was found to have a higher impact on the TPC compared with the TSC, that is, the liquid–solid ratio had the highest contribution (81.95%; $p < 0.001$), followed by water content (4.76%; $p < 0.001$) and ultrasonic power (3.79%; $p < 0.001$). In contrast, the extraction time had little influence on the TPC ($p > 0.05$). Zeng et al. [15] studied the effects of the water content added to the DES, extraction temperature, and liquid–solid ratios on the extraction yield of polyphenols through response surface analysis, and found that the water content and the liquid–solid ratio had the greatest impact on the extraction yield; this is consistent with the results of this study. Based on analyzing the recovery of saponin and polyphenol using the Taguchi method in this study, the optimal responses were found to be as follows: liquid–solid ratio of 40 mL/g, 30% water content in DES, 30 min of extraction time, and 300 W of ultrasonic power.

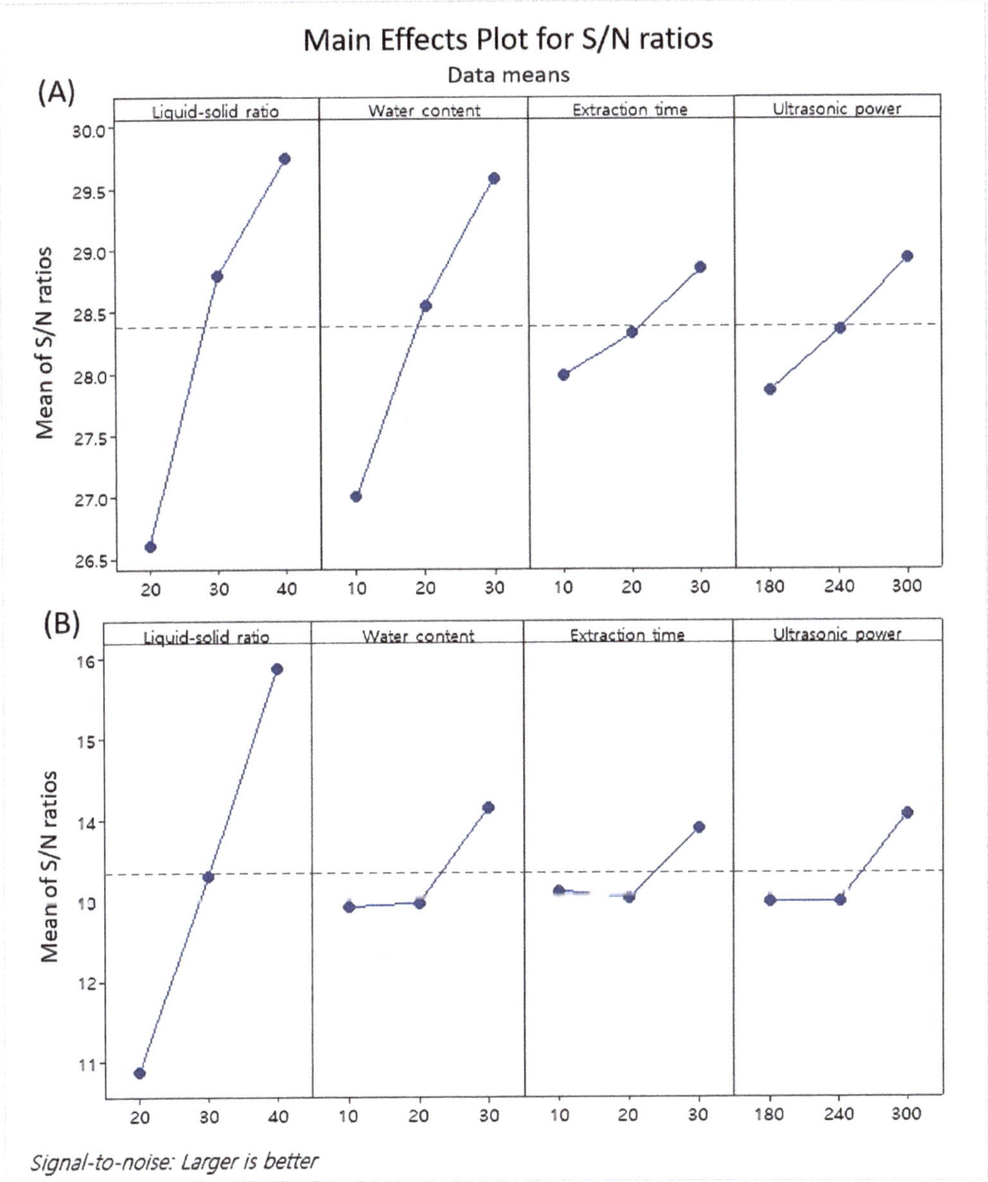

Figure 1. Main effects plot for mean of total saponin content (**A**) and total polyphenol content (**B**).

Table 2. Analysis of variance results of orthogonal design experiment on different extraction conditions of maca leaf saponins.

Factor	Degree of Freedom	Sum of Square	Mean of Square	F-Ratio	*p*-Value	Contribution (%)
Total saponin content						
Liquid–solid ratio	2	775.325	387.662	111.814	<0.001	51.85
Water content	2	494.352	247.176	71.294	<0.001	33.06
Extraction time	2	23.620	11.810	3.406	0.042	1.58
Ultrasonic power	2	46.080	23.040	6.646	0.003	3.08
Error	45	156.016	3.467			10.43
Total	53	1495.393				100
R^2	98.24%					
R^2 (adj)	96.48%					
Total polyphenol content						
Liquid–solid ratio	2	65.018	32.509	218.623	<0.001	81.95
Water content	2	3.780	1.890	12.710	<0.001	4.76
Extraction time	2	0.837	0.418	2.813	0.071	1.06
Ultrasonic power	2	3.007	1.504	10.111	<0.001	3.79
Error	45	6.692	0.149			8.44
Total	53	79.334				100
R^2	96.89%					
R^2(adj)	97.78%					

Contribution (%) = $\frac{Sum\ of\ square}{Total\ sum\ of\ square} \times 100$

Based on the experimental data presented in Table 1, linear regression analysis was performed to predict the relationship between the TSC or TPC and factors. The predicted equations, automatically formulated with Minitab 18 at a 95% confidence interval, were shown in Equations (1) and (2):

$$TSC = -0.25 + 0.4588A + 0.03695B + 0.0787C + 0.09839D \quad (1)$$

$$TPC = -1.005 + 0.1338A + 0.0283B + 0.0117 + 0.00425D \quad (2)$$

where A = liquid–solid ratio (mL/g), B = water content (%), C = extraction time (min), and D = ultrasonic power (W).

The *p* values of the derived regression equation for the TSC and TPC were 0.001 and 0.003, respectively, indicating the suitability of the proposed models (R^2 = 98.24%, adjusted R^2 = 96.48% for TSC; R^2 = 96.89%, adjusted R^2 = 93.78% for TPC).

2.4. Confirmation Test Results

In order to validate the predicted values derived from regression Equations (1) and (2) through the Taguchi method, the experimental values obtained under the optimal extraction conditions were measured and compared, and the results are presented in Table 3. Under optimal conditions, i.e., a liquid–solid ratio of 40 mL/g, water content of 30%, extraction time of 30 min, and ultrasonic power of 300 W, the predicted values for the TSC and TPC were 37.07 mg OAE/g DW and 6.82 mg GAE/g DW, respectively. The experimentally determined TSC and TPC values under the same extraction conditions were 38.02 mg OAE/g DW and 6.78 mg GAE/g DW, respectively. The differences between the predicted and experimental TSC and TPC values were minimal, at 0.95 and 0.04, respectively; this supported the notion that the regression model equations were suitably accurate within the 95% confidence interval, which ranged between 34.64 and 39.49 mg OAE/g DW for the TSC and 6.07 and 7.57 mg GAE/g DW for the TPC. Consequently, the close alignment between the experimental and predicted results, with the experimental values falling within the 95%

confidence interval, suggests that the models are robust and reliable for predicting the TSC and TPC of the extracts obtained under optimal extraction conditions.

Table 3. Confidence intervals and results of conformation experiments at optimal extraction conditions.

Response	Predicted Value	Predicted Confidence Interval at 95% Confidence Level	Experimental Value	Difference
Total saponin content (mg OAE/g DW)	37.07	34.64–39.49	38.02 ± 0.45	0.95
Total polyphenol content (mg GAE/g DW)	6.82	6.07–7.57	6.78 ± 0.12	0.04

OAE, oleanolic acid equivalent; GAE, gallic acid equivalent; and DW, sample dry weight. Experimental value is represented as mean ± standard deviation of triplicate.

2.5. Comparison of Bioactive Properties of Maca Leaf Extracts

To gain insight into the superiority of DES-based UAE for the bioactive compounds of maca leaves, the TSC, TPC, and antioxidant activity of extracts obtained via DES-based UAE were compared with those of HWE, a conventional extraction method, and the results are shown in Table 4. The TSC of the extract obtained via DES-based UAE was 718.89 mg OAE/g and that via HWE was 707.50 mg OAE/g, which were much higher than those of Lee and Chang [5], who reported that the TSC of the methanol extract of maca leaves was 6.09 mg escin equivalents/100 g DW. In particular, this study confirmed that DES-based UAE was more effective than HWE in extracting saponins. These results are considered to be because saponins are polar compounds that are soluble in polar solvents such as water and alcohol, but insoluble in nonpolar organic solvents such as chloroform, ether, and acetone [25]. Yang et al. [26] reported that the extraction yield of physiologically active steroid saponins from extracts obtained from Dioscoreae nipponicae rhizoma using choline chloride and malonic acid as solvents was higher than that of water extracts, which was consistent with the results of this study.

Table 4. Comparison of bioactive compound contents and antioxidant activity of maca leaf extracts obtained via deep eutectic solvent-based ultrasound-assisted extraction and hot water extraction.

		DES-Based UAE	HWE
Total saponin content (mg OAE/g)		718.89 ± 0.95 *	707.50 ± 2.18
Total polyphenol content (mg GAE/g)		260.19 ± 1.00 *	224.90 ± 1.65
Phenolic compounds (mg/g)	Catechin	102.58 ± 1.93 *	88.73 ± 0.25
	Epicatechin	7.96 ± 0.13 *	7.29 ± 0.14
	Gallic acid	17.82 ± 0.32 *	15.77 ± 0.25
	Cinnamic acid	8.24 ± 0.05 *	6.89 ± 0.10
	Ferulic acid	6.96 ± 0.04 *	5.25 ± 0.05
	Caffeic acid	3.89 ± 0.06 *	3.61 ± 0.15
	p-Coumaric acid	1.22 ± 0.02 *	0.97 ± 0.06
	Protocatechuic acid	6.92 ± 0.03 *	5.43 ± 0.21
	Naringin	4.78 ± 0.06 *	3.28 ± 0.11
IC_{50} value (mg/mL)	DPPH radical scavenging activity	0.95 ± 0.01 *	1.15 ± 0.03
	ABTS radical scavenging activity	0.83 ± 0.04 *	1.82 ± 0.03

DES, deep eutectic solvent; UAE, ultrasound-assisted extraction; HWE, hot water extraction; OAE, oleanolic acid equivalent; and GAE, gallic acid equivalent. Experimental value is represented as mean ± standard deviation of triplicate. * Significant differences between DES-based UAE and HWE extracts according to *t*-test ($p < 0.05$).

The TPC of the maca leaf extracts obtained via DES-based UAE and HWE were 260.19 mg GAE/g and 224.90 mg GAE/g, indicating that DES-based UAE is more effective in extracting polyphenols. In addition, when measuring the content of phenolic compounds in the extracts, the extract obtained via DES-UAE showed a higher content than the extract obtained via HWE. For example, the phenolic compound content of the extract obtained via DES-based UAE was the highest in catechin at 102.58 mg/g, followed by gallic acid (17.82 mg/g), cinnamic acid (8.24 mg/g), and protocatechuic acid (6.92 mg/g). The extract obtained via HWE showed the highest in catechin at 88.73 mg/g, followed by gallic acid (15.77 mg/g), epicatechin (7.29 mg/g), and cinnamic acid (6.89 mg/g). Compared with conventional extraction solvents, the high viscosity of DES has a limitation: the extraction efficiency can be reduced because slow mass transfer occurs. Therefore, the high phenolic compound contents of the extracts obtained via DES-based UAE is considered to be due to the fact that the addition of water to DES lowered the viscosity, resulting in high diffusivity, and thus improved the extraction efficiency [27].

The IC_{50} values for the 1,1-diphenyl-2-picryl hydrazyl (DPPH) and 2,2′-azino-bis(3-ethylbenzothiazoline-6-sulfonic acid (ABTS) radical scavenging activities were 0.95 and 0.83 mg/mL in DES-based UAE and 1.31 and 1.82 mg/mL in HWE, respectively, indicating that the extract obtained via DES-based UAE had excellent antioxidant activity. Campos et al. [28] analyzed the phenolic compounds in an ethanol extract of maca root and reported that gallocatechin derivatives were present in the highest concentrations, ranging between 2.5 and 20.7 mg GAE/100 mL, followed by catechin derivatives, protocatechuic acid, and *p*-coumaric acid derivatives, which were between 1.4 and 15.2 and 0.6 and 0.4 mg GAE/100 mL, respectively. In addition, Sandoval et al. [29] reported that the catechin content in the water extract of maca root was 2.5 mg/g, so the results of this study showed that the catechin content of the maca leaf extract obtained using DES-based UAE was significantly higher. Catechin, a flavonol found in various foods such as wine, tea, and fruits, is recognized for its hydroxyl and peroxyl radicals, superoxide, DPPH radical scavenging activities, and iron-chelating effects [30]. Furthermore, gallic acid and its derivatives, which are widely distributed in the plant kingdom, exhibit various physiological activities, including neuroprotective effects, free radical scavenging, and anti-cancer properties [31,32]. The results above demonstrate that a combination of DES and UAE enhances the extraction yield of bioactive compounds such as saponins and polyphenols from maca leaves and increases the antioxidant activity of the extract. This is thought to be because DES-based UAE operates under milder conditions, requires less solvent, and improves mass transfer due to cell wall destruction via ultrasound. Therefore, DES-based UAE is considered to be an environmentally friendly and energy-efficient extraction method compared to conventional HWE and could be applied to extract heat-sensitive bioactive components from various plants.

3. Material and Methods

3.1. Plant Materials and Chemical Reagents

The maca plants used in this study were procured from the Youth Research Institute (Cheongdo, Republic of Korea). After removing the roots and washing the leaves, they were freeze-dried. The dried leaves were then pulverized using a food mixer (FM-681C, Hanil Electric Co., Ltd., Seoul, Republic of Korea) and passed through a 45-mesh screen (Chung Gye Indus, MFG, Co., Seoul, Republic of Korea). The resulting leaf powder was stored in a deep freezer (MDF, Sanyo Electric Co., Ltd., Osaka, Japan) at −40 °C until use.

Diaion HP resin, catechin, epicatechin, gallic acid, cinnamic acid, ferulic acid, caffeic acid, *p*-coumaric acid, protocatechuic acid, naringin, phosphoric acid, DPPH, and ABTS were purchased from Sigma-Aldrich (St. Louis, MO, USA). A Folin–Ciocâlteu phenol reagent and oleanolic acid were purchased from Junsei (Tokyo, Japan), and 100% acetonitrile was purchased from Burdick & Jackson™ (Honeywell International Inc., Charlotte, NC, USA). Choline chloride and glycerin was purchased from Daejung (Siheung, Korea), and all other chemicals used were of analytical grade.

3.2. *Deep Eutectic Solvent Preparation*

The DES for extracting bioactive compounds from maca leaves was prepared based on the results of a preliminary experiment conducted by referencing the method of Barbieri et al. [33] and Cai et al. [34]. That is, choline chloride and glycerin were mixed in a molar ratio of 1:2 and then stirred in a shaking water bath (BS-11, JeioTech Co., Ltd., Seoul, Republic of Korea) at 120 rpm at 80 °C until the color became transparent. After cooling the DES, its stability was checked by leaving it at room temperature overnight.

3.3. *Optimization of Extraction Conditions*

3.3.1. Optimization Parameters Using Orthogonal Experiment Design

The optimization of DES-based UAE to extract the bioactive compounds from maca leaves was conducted according to the orthogonal experiment design of the Taguchi method [16]. Orthogonal matrix L_9 (3^4) was used, and four factors, liquid–solid ratio, water content added to DES, extraction time, and ultrasonic power, and three levels were set (Table 1). The extraction temperature was maintained at 38 $\pm$ 2 °C. To prevent an excessive increase in the extract temperature during extraction, a beaker containing the extract was placed in an ice bath and monitored using a temperature-sensing device. The DES-based UAE was performed according to each condition, and the response variables were TSC and TPC.

By applying Taguchi's design to the experimental data, main effects plots were generated to identify the significant factors and optimize their levels to maximize the recovery of bioactive compounds. The S/N ratios were analyzed to determine the optimal conditions for each parameter, with the 'larger is better' criterion selected to maximize the extraction yield. Additionally, ANOVA was conducted to assess the influence of each parameter on the extraction responses and to evaluate the general linear regression model. The Taguchi experimental design and ANOVA were performed using Minitab® 18 statistical software (Minitab Inc., State College, PA, USA).

3.3.2. Determination of Total Saponin Content

The TSC of maca leaf extracts was determined using the vanillin-sulfuric acid method, as described by Lee and Chang [5]. For each concentration, 0.25 mL of the prepared extract sample was mixed with 0.25 mL of 8% (*w*/*v*) vanillin solution (in ethanol) and 2.5 mL of 72% sulfuric acid. The mixture was then incubated in a water bath at 60 °C for 10 min, followed by cooling in cold water for 5 min. Then, the absorbance was measured at 544 nm using a microplate reader (Epoch, BioTek Instrument Inc., Winooski, VT, USA). The TSC was calculated using a standard curve created with oleanolic acid solutions at different concentrations. The TSC was expressed as mg of OAE per gram of sample DW.

3.3.3. Determination of Total Polyphenol Content

The TPC of the extract prepared from the maca leaves was determined using the method of Folin and Denis [35]. First, different concentrations of the extract were prepared. Then, 0.2 mL of Folin–Ciocâlteu's phenol reagent was added to the test solutions, mixed thoroughly, and they were allowed to stand at room temperature for 3 min. Subsequently, 0.4 mL of 10% sodium carbonate (Na_2CO_3) and 4 mL of distilled water were added to the mixture, which was then left in a dark room at room temperature for 1 h. Then, the absorbance was measured at 720 nm using a microplate reader (Epoch, Bioteck Instrument

Inc.). The TPC was quantified using a standard curve constructed with gallic acid solutions, and was expressed as mg GAE per gram of sample DW.

3.4. Comparison of Bioactive Properties of Maca Leaf Extracts

3.4.1. Preparation of Maca Leaf Extracts

To obtain biocomponent-rich extracts from maca leaves, DES-based UAE was performed under the optimal conditions selected through the Taguchi method. That is, 5 g of maca leaves were mixed with 200 mL of DES containing 30% water, and extracted using an ultrasonicator (KFS-600N, Korprotech, Seoul, Korea) for 30 min. The ultrasonic power was 300 W, and the extraction temperature was maintained at 38 $\pm$ 2 °C. The extract was centrifuged at 4 °C and 9896$\times$ *g* for 20 min (Supra-21K, Hanil, Daejeon, Korea), and the supernatant was filtered to remove the DES with Diaion HP resin. The distilled water (9.5 mL) was added to a glass column (20 $\times$ 200 mm) equipped with a filter, and 25 mL of Diaion HP 20 resin was filled and left until the resin hardened. After 5 mL of the maca leaf extract was loaded, DES was removed with 30 mL of distilled water. Afterwards, 50 mL of 100% ethanol was added to elute the bioactive compounds. The elution process was repeated three times, and the collected fractions were freeze-dried and used as an extract for measuring bioactive properties.

To verify the efficiency of the optimized DES-based UAE approach for the extraction of bioactive compounds, HWE, a representative conventional extraction method, was performed for comparative purposes. That is, the sample (5 g) was mixed with 200 mL of distilled water and extracted using a shaking water bath (BS-11, JeioTech Daejeon, Korea) at 90 °C for 2 h. The supernatant obtained via centrifugation was freeze-dried and used as a comparative sample to measure the functional component content and antioxidant activity.

3.4.2. Determination of Total Saponin and Polyphenol Contents

The TSC and TPC of maca leaf extracts obtained under optimal conditions were determined using the methods of Lee and Chang [5] and Folin and Denis [32], respectively, as described above. The TSC was expressed as mg of OAE and GAE per gram of extract powder, and TPC was expressed as mg GAE per gram of extract powder.

3.4.3. Determination of Phenolic Compounds via HPLC

The phenolic compound content of maca leaf extracts obtained under optimized DES-based UAE conditions was measured using a high-performance liquid chromatography (HPLC) system (Waters 2695, Waters Corp., Milford, MA, USA) according to the method of Nour et al. [36], with modifications. The HPLC detection system utilized an ultraviolet detector (Waters 2489, Waters Corp.). The analytical column used was an Atlantis® dC18 (4.6 $\times$ 150 mm, 5 µm, Waters Corp.); the column temperature was maintained at 34 °C. The mobile phase consisted of a gradient of 1% phosphoric and 100% acetonitrile. The injection volume was 10 µL, and the detection wavelength was set at 280 nm. The standards used for the calibration curve included catechin, epicatechin, gallic acid, cinnamic acid, ferulic acid, caffeic acid, p-coumaric acid, protocatechuic acid, and naringin. The quantity of the phenolic compounds in the extracts was calculated using the area under the calibration curve's standard peaks.

3.4.4. Measurement of Antioxidant Activity

The antioxidant activities of the maca leaf extracts obtained under optimal conditions were assessed using the DPPH and ABTS assays. Antioxidant activity was expressed as the half-maximal inhibitory concentration (IC_{50}) values for the DPPH and ABTS radical scavenging activities.

The DPPH radical scavenging activity was determined using the method of Lee and Yoon [37], with modifications. An extract sample (100 μL) and 0.2 mM DPPH (200 μL) solution were added to a microplate, incubated at 37 °C for 30 min, and the absorbance was measured at 517 nm using a microplate reader (Epoch, BioTek Instruments Inc., Winooski, VT, USA). The DPPH radical scavenging ability was calculated using the following Equation (3):

$$\text{DPPH radical scavenging activity (\%)} = (1 - (A - B)/C) \times 100 \quad (3)$$

where A is the absorbance of the test sample, B is the absorbance of distilled water instead of DPPH, and C is the absorbance of distilled water instead of the test sample.

The ABTS radical scavenging activity was measured using the method of Lee and Yoon [38]. Initially, 7 mM ABTS and 2.45 mM potassium persulfate were dissolved in distilled water and kept in the dark for 12–14 h to form ABTS cation radicals (ABTS+). The resulting solution was then diluted with 80% ethanol to achieve an absorbance of 0.700 ± 0.002 at 734 nm. Subsequently, 15 μL of extract sample and 300 μL of diluted ABTS+ solution were added to a microplate and allowed to react for 6 min. The absorbance of the mixture was measured at 734 nm using a microplate reader (Epoch, BioTek Instruments Inc.) and the ABTS radical scavenging activity was calculated using Equation (4).

$$\text{ABTS radical scavenging activity (\%)} = (1 - (A - B)/C) \times 100 \quad (4)$$

where A is the absorbance of the test sample, B is the absorbance of distilled water instead of ABTS, and C is the absorbance of distilled water instead of the test sample.

3.5. Statistical Analysis for Validation

The validation experiments performed to confirm the extraction efficiency of the DES-based UAE method were carried out in triplicate, and the experimental results were expressed as the mean ± standard deviation. Significant differences between the DES-based UAE and HWE at a significance level of $p < 0.05$ were analyzed Via a *t*-test using SPSS (Ver. 23, IBM Corp., Armonk, NY, USA).

4. Conclusions

This study aimed to use Taguchi's methodology to determine the optimal conditions for the DES-based UAE of bioactive compounds from maca leaves. Based on nine Taguchi design experiments, the optimal conditions for extracting saponins and polyphenols were determined as follows: 40 mL/g liquid–solid ratio, 30% water content in DES, 30 min extraction time, and 300 W ultrasonic power. Under optimized DES-based UAE conditions, the predicted values were 37.07 mg OAE/g and 6.82 mg GAE/g for the total saponin and polyphenol contents, respectively. The difference between the predicted and experimental values was minimal, indicating that the model was robust. Maca leaf extracts obtained under the optimized DES-based UAE conditions had higher saponins, polyphenols, phenolic compound contents, and antioxidant activities than were obtained Via HWE. In conclusion, DES-based UAE can be used as an efficient and economical alternative method for extracting bioactive compounds from maca leaves. Furthermore, the maca leaf extract obtained Via DES-based UAE offers potential applications as a functional food.

Author Contributions: Software, E.J.L.; formal analysis, E.J.L.; investigation, E.J.L.; resources, E.J.L.; writing—original draft preparation, E.J.L.; conceptualization, K.Y.Y.; methodology, K.Y.Y.; validation, K.Y.Y.; writing-review and editing, K.Y.Y.; supervision, K.Y.Y.; funding acquisition, K.Y.Y.; All authors have read and agreed to the published version of the manuscript.

Funding: This research was funded by the Yeungnam University [224A061023].

Institutional Review Board Statement: Not applicable.

Informed Consent Statement: Not applicable.

Data Availability Statement: The data are contained within the article.

Conflicts of Interest: The authors declare no conflicts of interest.

References

1. Quiros, C.F.; Epperson, A.; Hu, J.; Holle, M. Physiological Studies and Determination of Chromosome Number in Maca, *Lepidium meyenii* (Brassicaceae). *Econ. Bot.* **1996**, *50*, 216–223. [CrossRef]
2. del Carpio, N.U.; Alvarado-Corella, D.; Quiñones-Laveriano, D.M.; Araya-Sibaja, A.; Vega-Baudrit, J.; Monagas-Juan, M.; Navarro-Hoyos, M.; Villar-López, M. Exploring the Chemical and Pharmacological Variability of *Lepidium meyenii*: A Comprehensive Review of the Effects of Maca. *Front. Pharmacol.* **2024**, *15*, 1360422.
3. Wang, S.; Zhu, F. Chemical Composition and Health Effects of Maca (*Lepidium meyenii*). *Food Chem.* **2019**, *288*, 422–443. [CrossRef] [PubMed]
4. Lee, Y.K.; Chang, Y.H. Process Optimization of Methanol Extract from Maca Leaves and Its Physicochemical Properties. *J. Korean Soc. Food Sci. Nutr.* **2018**, *47*, 543–549. [CrossRef]
5. Lee, Y.K.; Chang, Y.H. Physicochemical and Antioxidant Properties of Methanol Extract from Maca (*Lepidium meyenii* Walp.) Leaves and Roots. *Food Sci. Technol.* **2019**, *39*, 278–286.
6. Lee, Y.K.; Chang, Y.H. Microencapsulation of a Maca Leaf Polyphenol Extract in Mixture of Maltodextrin and Neutral Polysaccharides Extracted from Maca Roots. *Int. J. Bio. Macromol.* **2020**, *150*, 546–558.
7. Jin, W.; Chen, X.; Huo, Q.; Cui, Y.; Yu, Z.; Yu, L. Aerial Parts of Maca (*Lepidium meyenii* Walp.) as Functional Vegetables with Gastrointestinal Prokinetic Efficacy in vivo. *Food Funct.* **2018**, *9*, 3456–3465. [CrossRef]
8. Putnik, P.; Kovačević, D.B.; Jambrak, A.R.; Barba, F.J.; Cravotto, G.; Binello, A.; Lorenzo, J.M.; Shpigelman, A. Innovative "Green" and Novel Strategies for the Extraction of Bioactive Added Value Compounds from Citrus Wastes—A Review. *Molecules* **2017**, *22*, 680. [CrossRef]
9. Kumar, K.; Srivastav, S.; Sharanagat, V. Ultrasound Assisted Extraction (UAE) of Bioactive Compounds from Fruit and Vegetable Processing By-products: A Review. *Ulrason. Sonochem.* **2020**, *70*, 105325. [CrossRef]
10. Zdanowicz, M.; Wilpiszewska, K.; Spychaj, T. Deep Eutectic Solvents for Polysaccharides Processing. A Review. *Carbohydr. Polym.* **2018**, *200*, 361–380. [CrossRef]
11. Cunha, S.C.; Fernandes, J.O. Extraction Techniques with Deep Eutectic Solvents. *TrAC, Trends Anal. Chem.* **2018**, *105*, 225–239. [CrossRef]
12. Zhang, L.; Wang, M. Optimization of Deep Eutectic Solvent-Based Ultrasound-Assisted Extraction of Polysaccharides from *Dioscorea opposite* Thunb. *Int. J. Biol. Macromol.* **2017**, *95*, 675–681. [PubMed]
13. Tang, W.; Wu, Y.; Wang, M.; Row, K.H.; Qiu, H.; Zhou, J.L. Emerging Application of Extraction Phase of Ionic and Non-ionic Deep Eutectic Solvents Toward Natural Herbal Medicine. *TrAC Trends Anal. Chem.* **2023**, *165*, 117137.
14. Ali, M.C.; Chen, J.; Zhang, H.; Li, Z.; Zhao, L.; Qiu, H. Effective Extraction of Flavonoids from *Lycium barbarum* L. Fruits by Deep Eutectic Solvents-Based Ultrasound-Assisted Extraction. *Talanta* **2019**, *203*, 16–22. [CrossRef] [PubMed]
15. Zeng, J.; Dou, Y.; Yan, N.; Li, N.; Zhang, H.; Tan, J. Optimizing Ultrasound-Assisted Deep Eutectic Solvent Extraction of Bioactive Compounds from Chinese Wild Rice. *Molecules* **2019**, *24*, 2718. [CrossRef]
16. Kim, E.J.; Kim, C.Y.; Yoon, K.Y. Application of Taguchi Method to Optimize Deep Eutectic Solvent-Based Ultrasound-Assisted Extraction of Polysaccharides from Maca and Its Biological Activity. *Food Bioproc. Tech.* **2025**, *18*, 27092720.
17. Liu, Y.; Li, Z.; Xu, H.; Han, Y. Extraction of Saponin from *Camellia oleifera* Abel Cake by a Combination Method of Alkali Solution and Acid Isolation. *J. Chem.* **2016**, *2016*, 1–8.
18. Salacheep, S.; Kasemsiri, P.; Pongsa, U.; Okhawili, M.; Chindaprasirt, P.; Hiziroglu, S. Optimization of Ultrasound-Assisted Extraction of Anthocyanins and Bioactive Compounds from Butterfly Pea Petals using Taguchi Method and Grey Relational Analysis. *J. Food Sci. Technol.* **2020**, *57*, 3720–3730.
19. Lu, W.; Liu, S. Choline Chloride–Based Deep Eutectic Solvents (Ch-DESs) as Promising Green Solvents for Phenolic Compounds Extraction from Bioresources: State-of-the-Art, Prospects, and Challenges. *Biomass Convers. Biorefin.* **2022**, *12*, 2949–2962. [CrossRef]
20. Duan, L.; Dou, L.; Guo, L.; Li, P.; Liu, E. Comprehensive Evaluation of Deep Eutectic Solvents in Extraction of Bioactive Natural Products. *ACS Sustain. Chem. Eng.* **2016**, *4*, 2405–2411.
21. Chen, Y.; Yu, D.; Chen, W.; Fu, L.; Mu, T. Water Absorption by Deep Eutectic Solvents. *Phy. Chem. Chem. Phys.* **2019**, *21*, 2601–2610.

22. Tang, Y.; He, X.; Sun, J.; Liu, G.; Li, C.; Li, L.; Sheng, J.; Zhou, Z.; Xin, M.; Ling, D. Comprehensive Evaluation on Tailor-made Deep Eutectic Solvents (DESs) in Extracting Tea Saponins from Seed Pomace of *Camellia oleifera*. *Abel. Food Chem.* **2020**, *342*, 128243. [PubMed]
23. Wang, Y.; Hou, Y.; Wu, W.; Liu, D.; Ji, Y.; Ren, S. Roles of a Hydrogen Bond Donor and a Hydrogen Bond Acceptor in the Etraction of Toluene from n-Heptane Using Deep Eutectic Solvents. *Green Chem.* **2016**, *18*, 3089–3097. [CrossRef]
24. Oztop, M.H.; Sahin, S.; Sumnu, G. Optimization of Microwave Frying of Potato Slices by Using Taguchi Technique. *J. Food Eng.* **2007**, *79*, 83–91. [CrossRef]
25. Negi, J.S.; Negi, P.S.; Pant, G.J.; Rawat, M.; Negi, S.K. Naturally Occurring Saponins: Chemistry and Biology. *J. Poisonous. Med. Plants. Res.* **2013**, *1*, 1–6.
26. Yang, G.Y.; Song, J.N.; Chang, Y.Q.; Wang, L.; Zheng, Y.G.; Zhang, D.; Guo, L. Natural Deep Eutectic Solvents for the Extraction of Bioactive Steroidal Saponins from Dioscoreae Nipponicae Rhizoma. *Molecules* **2021**, *26*, 2079.
27. Dai, Y.; Witkamp, G.J.; Verpoorte, R.; Choi, Y.H. Tailoring Properties of Natural Deep Eutectic Solvents with Water to Facilitate Their Applications. *Food Chem.* **2015**, *187*, 14–19.
28. Campos, D.; Chirinos, R.; Barreto, O.; Noratto, G.R. Optimized Methodology for the Simultaneous Extraction of Glucosinolates, Phenolic Compounds and Antioxidant Capacity from Maca (*Lepidium meyenii*). *Ind. Crops Prod.* **2013**, *49*, 747–754.
29. Sandoval, M.; Okuhama, N.N.; Angeles, F.M.; Melchor, V.V.; Condezo, L.A.; Lao, J.; Miller, M.J. Antioxidant Activity of the Cruciferous Vegetable Maca (*Lepidium meyenii*). *Food Chem.* **2002**, *79*, 207–213. [CrossRef]
30. Yilmaz, Y. Novel Uses of Catechins in Foods. *Trends Food Sci. Technol.* **2006**, *17*, 64–71.
31. Lu, Z.; Nie, G.; Belton, P.S.; Tang, H.; Zhao, B. Structure–Activity Relationship Analysis of Antioxidant Ability and Neuroprotective Effect of Gallic Acid Derivatives. *Neurochem. In.* **2006**, *48*, 263–274. [CrossRef] [PubMed]
32. Pal, S.M.; Avneet, G.; Siddhraj, S.S.; Asdaq, S. Gallic Acid: Pharmacogical Promising Lead Molecule: A Review. *Int. J. Pharmacogn. Pharm. Res.* **2018**, *10*, 132–138.
33. Barbieri, J.B.; Goltz, C.; Cavalheiro, F.B.; Toci, A.T.; Igarashi-Mafra, L.; Mafra, M.R. Deep Eutectic Solvents Applied in the Extraction and Stabilization of Rosemary (*Rosmarinus officinalis* L.) phenolic compounds. *Ind. Crops Prod.* **2020**, *144*, 112049. [CrossRef]
34. Cai, Y.Q.; Gao, H.; Song, L.M.; Tao, F.Y.; Ji, X.Y.; Yu, Y.; Cao, Y.Q.; Tang, S.J.; Yue, P. Optimization of Green Deep Eutectic Solvent (DES) Extraction of *Chenopodium quinoa* Willd. Husks Saponins by Response Surface Methodology and Their Antioxidant Activities. *RSC Adv.* **2023**, *13*, 29408. [CrossRef]
35. Folin, O.; Denis, W. On Phosphotungstic-Phosphomolybdic Compounds as Color Reagents. *J. Biol. Chem.* **1912**, *12*, 239–243. [CrossRef]
36. Nour, V.; Trandafir, I.; Cosmulescu, S. HPLC Determination of Phenolic Acids, Flavonoids and Juglone in Walnut Leaves. *J. Chromatogr. Sci.* **2012**, *51*, 883–890. [CrossRef]
37. Lee, S.H.; Yoon, K.Y. Physicochemical Properties, Bioactive Ingredients, and Antioxidant Activity of Cheonggukjang Added with Brewer's Spent Grain. *Food Sci. Preserv.* **2024**, *31*, 683–692. [CrossRef]
38. Lee, J.J.; Yoon, K.Y. Antioxidant, Antidiabetic and Anti-inflammatory Activities of Bitter Melon Extracts Obtained by Ultrasonic-Assisted Extraction. *Korean J. Food Preserv.* **2022**, *29*, 777–789. [CrossRef]

molecules

MDPI

Article

High-Power Ultrasound and High-Voltage Electrical Discharge-Assisted Extractions of Bioactive Compounds from Sugar Beet (*Beta vulgaris* L.) Waste: Electron Spin Resonance and Optical Emission Spectroscopy Analysis

Josipa Dukić [1], Anet Režek Jambrak [1,*], Jurica Jurec [2], Dalibor Merunka [2], Srećko Valić [2,3,4], Rafaela Radičić [5], Nikša Krstulović [5], Marinela Nutrizio [1] and Igor Dubrović [6]

[1] Faculty of Food Technology and Biotechnology, University of Zagreb, 10000 Zagreb, Croatia; josipa.dukic@pbf.unizg.hr (J.D.); marinela.nutrizio@pbf.unizg.hr (M.N.)
[2] Ruđer Bošković Institute, 10000 Zagreb, Croatia; jjurec@irb.hr (J.J.); dalibor.merunka@irb.hr (D.M.); srecko.valic@irb.hr (S.V.)
[3] Department of Medical Chemistry, Biochemistry and Clinical Chemistry, Faculty of Medicine, University of Rijeka, 51000 Rijeka, Croatia
[4] Centre for Micro- and Nanosciences and Technologies, University of Rijeka, 51000 Rijeka, Croatia
[5] Institute of Physics, Bijenička Cesta 46, 10000 Zagreb, Croatia; rradicic@ifs.hr (R.R.); niksak@ifs.hr (N.K.)
[6] Department of Environmental Health, Teaching Institute of Public Health of Primorje-Gorski Kotar County, 51000 Rijeka, Croatia; igor.dubrovic@zzjzpgz.hr
* Correspondence: anet.rezek.jambrak@pbf.unizg.hr

Abstract: To achieve sustainable extractions, this study examines the impact of different extraction methods to utilize waste from the sugar industry. In addition to conventional thermal extraction, the impact of high-power ultrasound (US) and high-voltage electrical discharge (HVED)-assisted extractions on the yield of bioactive compounds and the antioxidant capacity (AC) value of sugar beet leaf extracts was determined. US extraction proved to be an excellent method for extracting bioactive compounds, while HVED extraction proved to be an excellent method for extracting Vitexin. AC was measured both spectrophotometrically (DPPH and FRAP) and spectroscopically via electron spin resonance (ESR). The AC results correlate with each other, and the highest AC values were found in the US-treated samples with 25% ethanol solution as the extraction solvent. Characterization of the plasma via optical emission spectroscopy (OES) showed that neither the solvent nor the sample influenced the plasma spectra, only the gas used (nitrogen/argon). All of the obtained results provide an excellent basis for future research into the utilization of food waste and by-products.

Keywords: high-voltage electrical discharge; high-power ultrasound; sugar beet leaves; bioactive compounds; antioxidant capacity; electron spin resonance spectroscopy; optical emission spectroscopy

1. Introduction

In the face of global challenges such as resource depletion, environmental degradation, and the increasing food needs of a rapidly growing population, the principles of sustainable development are critical to modern food production and the stability of the food industry. At its core, sustainable development stands for responsible, intelligent and creative use of natural resources, for social policies that are fair to all people regardless of age, ethnicity and gender, and for long-term economic stability based on economic growth without

 https://doi.org/10.3390/molecules30040796

negative impacts on nature and the population. By accepting the Sustainable Development Goals (SDGs) and incorporating the defined practices, the opportunity is created to protect fragile ecosystems, reduce greenhouse gas emissions and preserve valuable resources for future generations. The concept of sustainable development is central to the future of food production, considering that with current trends, the population is expected to increase rapidly, most likely exceeding 9 billion people by early 2050, according to some estimates [1]. In the face of population growth, the food sector is also facing the severe consequences of climate change, leading to losses of raw materials for production and thus jeopardizing the security of the supply chain. Global food production is expected to increase by 70%, leading to increased demand for raw materials and energy [2]. Therefore, innovations in the food sector are increasingly focusing on new, more energy-efficient and environmentally friendly raw materials and processing methods, which are recognized as valuable tools to achieve the SDGs [3–6].

In modern food technology and production, the use of ultrasound (US) represents an extremely valuable technology that, in addition to its versatility, has enabled an increase in processing efficiency, an improvement in product quality and a reduction in energy consumption. Regardless of whether it is high- or low-intensity US, this technology is widely used in the food industry for quality control, production, hygiene and by-product/waste processing/extraction [7]. Previous studies have shown that some of the advantages of using US in the extraction of bioactive components are the reduced extraction time without loss of yield, the low process temperature which reduces the degradation of sensitive compounds, the lower energy and solvent consumption, the possibility of using water as a solvent and thus avoiding toxic waste, the possibility of applications in the extraction of different bioactive compounds, the simplicity of the process and the facilitated subsequent separation and purification of the extracted components [8]. The above-mentioned advantages of US are due to the mechanical and thermal effects of transient acoustic cavitation, which cause the destruction of cell membranes, increased mass transport and more thorough penetration of solvents into the intracellular matrix [9].

Due to the numerous shortcomings of traditional, conventional methods for the extraction of bioactive compounds, innovative "green" non-thermal extraction methods that are economical and environmentally friendly are increasingly being used. The use of non-thermal extraction methods reduces energy and solvent consumption, increases the efficiency and selectivity of the process, ensures yields are higher, and makes automation of the process possible. In addition to US-assisted extraction, non-thermal techniques also include high-voltage electrical discharge (HVED)-assisted extraction, a non-thermal technique in which plasma is generated. Plasma is one of the four fundamental states of matter, i.e., a fully or partially ionized gas consisting of cations, anions, radicals, electrons and excited and unexcited atoms [10]. Some of the practical applications of this method are water purification, inactivation of microorganisms (bacteria, viruses, yeasts), electrohydraulic comminution and extraction of bioactive compounds from plants [11]. There is a difference between thermal and non-thermal, i.e., cold plasmas. In thermal plasmas, the particles (electrons, ions and neutrons) are in thermal equilibrium, i.e., they have the same temperature. In cold plasmas, electrons have different electron and ion temperatures and are not in local thermal equilibrium with ions and neutrons. Electrons can collide with other present molecules (N_2, O_2 and H_2O), creating secondary electrons, photons, ions and radicals in the plasma [12]. Non-thermal cold plasma generated by corona and current discharges implies the creation of plasma without heating the medium. Although the temperature in the discharge channels can be very high, a moderate or room temperature can be maintained due to the small volume and local nature of the discharge channels [13].

The low temperature of the extractions makes these methods desirable for the extraction of bioactive components with antioxidant capacity. Antioxidants are molecules that are crucial for the proper functioning of cells and biochemical processes, as they provide protection against oxidative stress and negative reactions with free radicals. Phenolic compounds are a large group (flavonoids, phenolic acid, tannins, stilbenes and lignans) of natural antioxidants found in plants. They are usually biosynthesized via shikimic acid from phenylalanine or tyrosine as initial reactants, and the hydroxyl (OH) group in the benzene ring is responsible for their antioxidant capacity [14]. In addition to the presence, the position and number of OH groups are also of great importance for the biological activity of flavonoids [15]. Phenolic compounds belong to the group of primary antioxidants, which are characterized by high efficacy and easy regenerability [16]. The consumption of phenolic compounds from food has been shown to have positive effects on human health, such as anti-inflammatory, anti-cancer and antiviral effects; improvement of immune system function; prevention of gastrointestinal tract diseases; etc. [17]. In the face of modern challenges such as climate change, the increase in external stressors and the huge amounts of food waste, scientific research in the food industry has focused on the use of advanced green processing techniques in the extraction of phenolic compounds from already existing but equally valuable waste such as sugar beet leaves [18,19]. Considering that sugar beet leaves represent about 30% of the whole plant, it is important to consider the use of this waste as a potential raw material for sustainable production needs [20]. The beneficial nutrient profile of leaves does not only refer to the macronutrients they contain. In terms of micronutrients, plant leaves, including sugar beet leaves, are generally a good source of minerals (Ca, Cu, Fe, K, Mg, Mn), vitamins (A, C, E, K) and phytochemicals/secondary metabolites (chlorophylls, carotenes), some of which have been shown to have a positive impact on human health [20,21]. There are relatively few studies on the extraction and characterization of the bioactive components contained in sugar beet leaves, but in general, sugar beet leaves contain high amounts of gallic acid (344 μg/mL) and ferulic acid (89.7 μg/mL), while cinnamic and caffeic acids and catechin are present in small amounts [22]. In addition, vitexin was found to be the most abundant phenolic compound [23].

For the above reasons, this study evaluated the efficacy of US and HVED treatments on the extraction of bioactive components with antioxidant capacity from sugar beet leaves by varying the extraction solvent, applied amplitude and treatment time for US treatments and the extraction solvent, applied gas, applied voltage and treatment time for HVED treatments. The focus is on sustainability and the use of industrial waste as a source of bioactive compounds combined with a low-environmental-impact extraction method.

2. Results and Discussion

2.1. Total Phenolic Content (TPC)

The results of the TPC yield for US-treated samples with deionized water as an extraction solvent were already published by Dukić et al. [19], but the extractions were repeated due to the second batch of plant material, and the results of the TPC yield for all US-treated samples are presented graphically (Figure 1a). The trend of previously published results is the same now, with a slight decrease in TPC yield observed in the new batch of plant material.

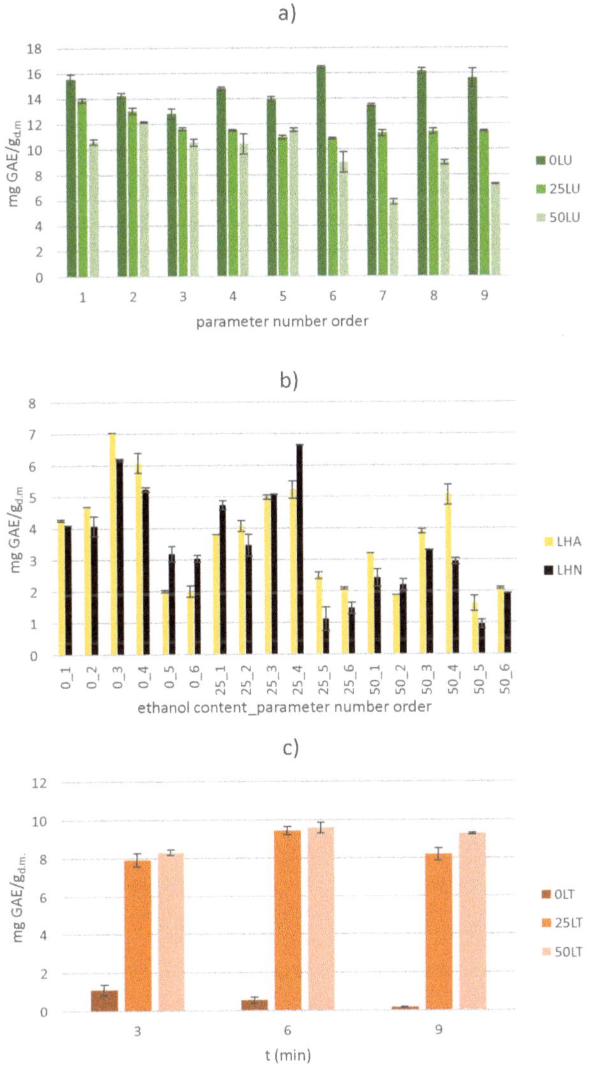

Figure 1. Determination of total phenolic content (TPC) for (**a**) US-treated samples, (**b**) HVED-treated samples and (**c**) thermally treated samples.

The reason for this decrease is due to the storage time and storage conditions of the plant material, which directly affect the chemical composition of the leaf itself. As in the previously mentioned work, the highest yield of TPC (16.47 ± 0.10 mg/$g_{d.m.}$) was determined to be that of the 0LU6 sample (with an amplitude of 100%, a treatment time of 9 min and deionized water as the extraction solvent). Maravić et al., 2022, reported a similar TPC yield for US-treated sugar beet leaf samples [23]. The increase in ethanol content (in the extraction solvent) directly affected the decrease in TPC yield in the US-treated samples, and thus the lowest TPC yield of 5.85 ± 0.21 mg/$g_{d.m.}$ was recorded for the 50LU7 sample (with an amplitude of 50%, a treatment time of 3 min and 50% ethanol solution as the extraction solvent). This is also confirmed by the statistical analysis of the results, where a statistically negative influence of ethanol content on the TPC yield was observed in the US-treated samples with a confidence interval of 95% ($p < 0.05$). A trend of decreasing TPC yields with increasing ethanol content was also observed in blackthorn fruit extracts [24], where TPC yields were 4.64–6.85 times lower when using a 100% ethanol solution as the extraction solvent than when using lower concentrations of ethanol solutions. In general, the TPC yields of US-treated samples are lower than the TPC yields obtained in the study by Ebrahimi et al. (2024) with the same plant material and using the same extraction method [25]. In the mentioned study, depending on the solvent used (0, 25, 50, 75 and 100% ethanol solution), the treatment time (4, 6 and 8 min), the amplitude used (25, 30 and 35%) and the ratio of plant material and solvent (1, 2, 3, 4 and 5% *w*/*v*), TPC yields between 2.05 $\pm$ 0.10 and 12.12 $\pm$ 0.09 mmol GAE/L (approx. 38.78 to 45.80 mg/$g_{d.m.}$) were obtained. Considering the ratio between the diameter of the US probe; the volume of the extraction mixture, which is almost twice as large in the aforementioned work (2 mm for 10 mL vs. 12 mm for 100 mL); and the shape of the laboratory dishes (Falcon Tubes of 15 mL vs. laboratory beakers of 250 mL), the yields obtained are lower, as expected. The diameter of the probe is one of the most important parameters that directly influences the efficiency of US extraction. In general, when using smaller diameter probes, the effect of transient acoustic cavitation is greater, but within a narrow range. In contrast, probes with a larger diameter distribute the energy over a larger area, so that the effect of the US is stronger.

Compared to the results of the US extraction, the yield of TPC after the HVED-assisted extraction was up to 17.34 times lower (Figure 1b). In particular, the highest yield of TPC in HVED-treated samples was observed in sample 0LHA3 (with deionized water as the extraction solvent, argon as the applied gas, an applied voltage of 20 kV and a treatment time of 9 min) and was 7.04 $\pm$ 0.00 mg GAE/$g_{d.m.}$, while the lowest amount was 0.95 $\pm$ 0.14 mg GAE/$g_{d.m.}$ for the 50LHN5 sample (with 50% ethanol solution as the extraction solvent, nitrogen as the applied gas, an applied voltage of 20 kV and a treatment time of 3 min). Although the highest and lowest yields were obtained using different gases, no statistically significant influence of the gases was found ($p > 0.05$). In contrast, treatment time, applied voltage and ethanol content showed a statistically significant impact on the TPC yield in the HVED-treated samples. In particular, a longer treatment time had a positive effect on the TPC yield, while the ethanol content and the applied voltage had a negative effect, which consequently led to a lower TPC yield. Comparing different treatment times (3, 6 and 9 min) at the same applied voltage (e.g., 20 kV), gas (e.g., argon) and ethanol content (e.g., deionized water as the extraction solvent), the positive effect of the treatment time was clearly visible. Specifically, a TPC yield of 2.01 $\pm$ 0.03 mg GAE/$g_{d.m.}$ was observed under the given conditions with a treatment duration of 3 min. A longer HVED treatment increased the TPC yield to 4.68 $\pm$ 0.00 mg GAE/$g_{d.m.}$ for a 6 min treatment, and 7.04 $\pm$ 0.00 mg GAE/$g_{d.m.}$ for a 9 min treatment. On the other hand, a higher ethanol content and an increase in the applied voltage from 20 kV to 25 kV had a negative effect on the TPC yield, regardless of the gas used. Some phenolic compounds are bound to the cell membrane, and it is necessary to invest a little more energy to extract them. For this reason, a longer treatment time has a positive effect on the yield, as a longer treatment time invests more energy and thus increases the chance that the bonds between the phenolic compound and the cell wall are broken and the phenolic compounds become more available for extraction. In contrast, a higher applied voltage causes the formation of a greater amount of molecular oxygen, which ultimately leads to the formation of molecular ozone, which negatively affects the phenolic compounds by breaking the bonds in their aromatic rings. The negative effect of increasing the voltage was also observed in samples of green tea powder, where a decrease in the TPC yield of 2.68–7.66 mg GAE/L was observed at the same treatment time but at a higher voltage (25 kV vs. 20 kV) [26].

In parallel to the non-thermal extraction methods, a conventional thermal extraction method at 60 °C was also performed and the obtained TPC yields are shown graphically

(Figure 1c). In general, the TPC yields obtained after thermal extraction with deionized water as the extraction solvent were significantly lower than the yields obtained with the same solvent but with non-thermal extraction methods (US and HVED). However, when ethanol solutions were used, higher yields were observed compared to the HVED-treated samples, but lower compared to US-treated samples. Ethanol content was a statistically significant variable of thermal extraction with a positive influence on the TPC yield ($p < 0.05$). This trend is opposite to the trends observed for US- and HVED-treated samples. The highest TPC yield of 9.58 $\pm$ 0.28 mg GAE/$g_{d.m.}$ was observed in the 50LT6 thermally treated sample (with 50% ethanol solution as the extraction solvent and a treatment time of 6 min), while the lowest yield of 0.16 $\pm$ 0.01 mg GAE/$g_{d.m.}$ was observed in the 0LT9 sample (with deionized water as the extraction solvent and a treatment time of 9 min). Considering only the 0LT thermally treated samples (with deionized water as the extraction solvent), it was found that the TPC yield decreased with longer treatments (up to 6.25 times). This is consistent with the heat sensitivity of phenolic compounds, which decompose or degrade at high temperatures and longer treatment times, leading to a decrease in TPC yield [27]. A decrease of 89% in the high-temperature treatment at 110 °C (compared to the treatment at 70 °C) was also observed in samples of myrtle leaves [28].

2.2. Vitexin Content

Regardless of the applied extraction method, the yield of vitexin decreases statistically significantly with increasing ethanol content ($p < 0.05$). Accordingly, the lowest vitexin yield was observed when a 50% ethanol solution was used as the extraction solvent. Specifically, a decrease in the vitexin yield of 2.77–5.02 times was observed in the US-treated samples, of 2.12–6.72 times for the HVED-treated samples and of 3.38–4.87 times for thermally treated samples. By optimizing the process parameters of the mentioned extractions and to achieve an optimal yield of vitexin, the ethanol content in the extraction solvent should be 0%. The obtained results correlate with the very poor solubility of the mentioned compound in ethanol and similar solvents. Due to its chemical structure (apigenin-8-C-glucoside), the flavonoid vitexin is highly soluble in organic solvents such as dimethyl sulfoxide (DMSO), but significantly less soluble in water and ethanol [29]. The results of vitexin yield for each extraction method are shown graphically (Figure 2).

The highest yield of vitexin in the thermally treated samples was 1369.91 $\pm$ 19.82 ng/mL, and was recorded in the 0TL3 sample (with deionized water as the extraction solvent and a treatment time of 3 min). Temperatures between 50 and 70 °C were found to be optimal temperatures for the extraction of vitexin from dried Mas Cotek leaves [30]. Namely, gentle heating has the ability to soften the plant tissue, weaken the integrity of the cell wall and consequently promote the release of bound active compounds from the plant material [27]. In order to achieve optimal yield, the optimal input parameters of thermal extraction were determined by statistical projection and corresponded to the parameters of the 0TL3 sample, which also showed the highest yield of vitexin. These parameters for thermal extraction are also the most acceptable from an energetic and environmental point of view, as the extraction takes the least time (and therefore the carbon footprint is the lowest) and the solvent used is also environmentally friendly.

Regarding the non-thermal extraction methods, the yield of vitexin was slightly higher in the HVED-treated samples than in the thermally treated samples. In particular, the highest yield was observed in the 0LHN3 sample (with deionized water as the extraction solvent, using nitrogen gas, an applied voltage of 20 kV and a treatment time of 9 min) and was 1481.22 $\pm$ 5.70 ng/mL. In addition to the extraction solvent, the applied voltage also had a statistically significant influence on the yield of vitexin in the HVED-treated samples ($p < 0.05$), while the application of different gases (argon/nitrogen) had no statistically significant influence on the yield of vitexin ($p > 0.05$). As well as a higher ethanol content, the application of higher voltages (25 kV) in the HVED-treated samples also had a statistically negative effect on the vitexin yield. To achieve the optimum values, the treatment time was statistically determined at 6.25 min, the ethanol content at 0%, the applied voltage at 20 kV and nitrogen gas was used. A decrease in the yield of flavonoids (by 558.54–2552.22 μg quercetin/mL) due to the application of a higher voltage (from 20 to 25 kV) was also observed in green tea powder samples [26]. In addition to TPC, higher voltage also has a negative effect on the yield of flavonoid compounds such as vitexin. The increased formation of molecular ozone and the breaking of bonds in their aromatic rings leads to a decrease in the yield of the flavonoid compound. Furthermore, exposure to higher voltages potentially leads to denaturation of the enzyme phenylalanine ammonialyse, an enzyme essential for the synthesis of flavonoid compounds [31].

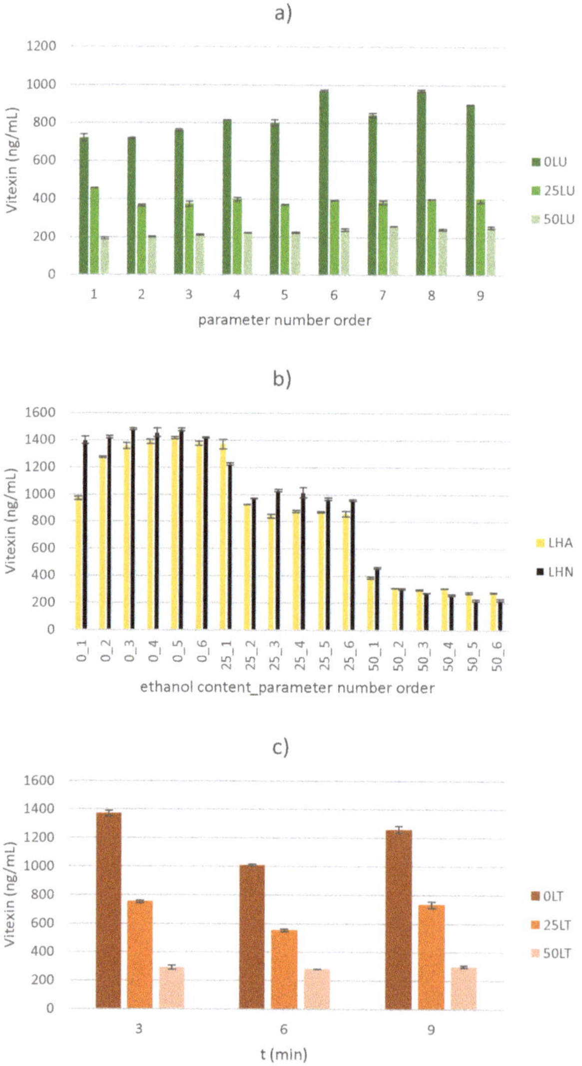

Figure 2. Determination of vitexin content for (**a**) US-treated samples, (**b**) HVED-treated samples and (**c**) thermally treated samples.

In contrast to the HVED-treated samples, the US-treated samples showed the lowest yield of vitexin. Although the obtained values are the lowest, they were not significantly different from the vitexin yield results obtained with other extraction methods, which ranged from 193.15 ± 5.60 ng/mL to 969.97 ± 3.52 ng/mL. The achieved yields are well below the yields determined in the study by Maravić et al., 2022 [23]. However, taking into account the ratio of plant material to solvent in the aforementioned study (1:10 compared to 1:50 in this work) as well as the treatment time of 30 min (and up to ten times longer) and the temperature of 50 °C (and up to 20 °C higher), such differences in yields can be explained [23]. Consequently, a longer treatment time with a higher concentration of plant material and the application of a higher temperature leads to a stronger formation of broken bonds between flavonoids and sugars, making the vitexin more available for the solvent. In this work, the yield of vitexin is statistically influenced by the input variables amplitude and the mutual interaction of amplitude and ethanol content ($p < 0.05$). In contrast to the statistically negative influence of the ethanol content and the mutual negative interaction between the applied amplitude and the ethanol content, the input variable amplitude has a positive effect on the vitexin yield. Specifically, by applying the same solvent for the same treatment time, increasing the amplitude increases the yield of the desired component, i.e., vitexin. The trend of a statistically significant increase in vitexin yield with increasing amplitude was also observed in samples of *Trollius chinensis* flowers [32]. As the US amplitude increases, the energy of the US wave increases, making the phenomenon of transient acoustic cavitation more prominent. Bubbles in the liquid medium implode and subsequently lead to large shear forces in the liquid medium, turbulence and mixing of the sample. Among other things, this leads to sonoporation and fragmentation of the plant material, making the desired components (e.g., vitexin) more available for the solvent and thus for extraction [33–35].

2.3. *Determination of Antioxidant Capacity (AC)*

2.3.1. Ferric Reducing Antioxidant Power (FRAP) Assay

One of the methods for measuring AC is the FRAP method, a very fast, cheap and simple spectrophotometric method that does not require any special equipment. However, the results obtained do not represent "real" values, as antioxidants, such as some proteins and carotenoids, which act on the principle of hydrogen transfer, cannot be determined [36,37]. In addition, antioxidants that contain a thiol group (-SH) in their structure, such as glutathione, cannot be detected using this method either [38]. Since the reaction itself requires a redox potential lower than the Fe^{3+}/Fe^{2+} redox potential, many compounds (not necessarily antioxidants) react with a lower redox potential, resulting in false high positive values of AC [39]. Regardless of the redox potential, some compounds still absorb at the determination wavelength due to their chemical composition, which at the same time has no antioxidant properties, which also leads to false positive values of AC [40]. On the other hand, some compounds such as quercetin, tannic acid, caffeic acid and ferulic acid react more slowly, so they are not able to reduce Fe^{3+} due to the shorter reaction time, so the results obtained are lower than the actual AC values [41]. Despite these shortcomings, the AC of the samples was measured and the results obtained are presented graphically (Figure 3).

The highest AC values were found in the US-treated samples, where a range of 1677.16 ± 36.98 to 2290.50 ± 56.69 μM Fe^{2+} was observed depending on the applied amplitude, extraction solvent and treatment time. Although no input parameter had a statistically significant effect ($p > 0.05$) on the CA value of the US-treated samples, the highest values were observed when a 25% ethanol solution (extraction solvent) was used. The highest value was recorded for the 25LU3 sample, where the AC value of 2290.50 ± 56.69 μM Fe^{2+} was obtained with 25% ethanol solution as the extraction solvent at an amplitude of

50% over a treatment time of 6 min. The application of higher amplitudes may lead to the formation of smaller air bubbles which directly affect the extraction efficiency and reduce the efficiency due to the lower transfer of ultrasonic energy to the solvent [42]. Although it did not statistically significantly affect the AC value of the US-treated samples, treatment time is an important extraction factor. A longer extraction time can lead to a higher yield of compounds with AC, but too long treatment has a negative effect on it by degrading it [43].

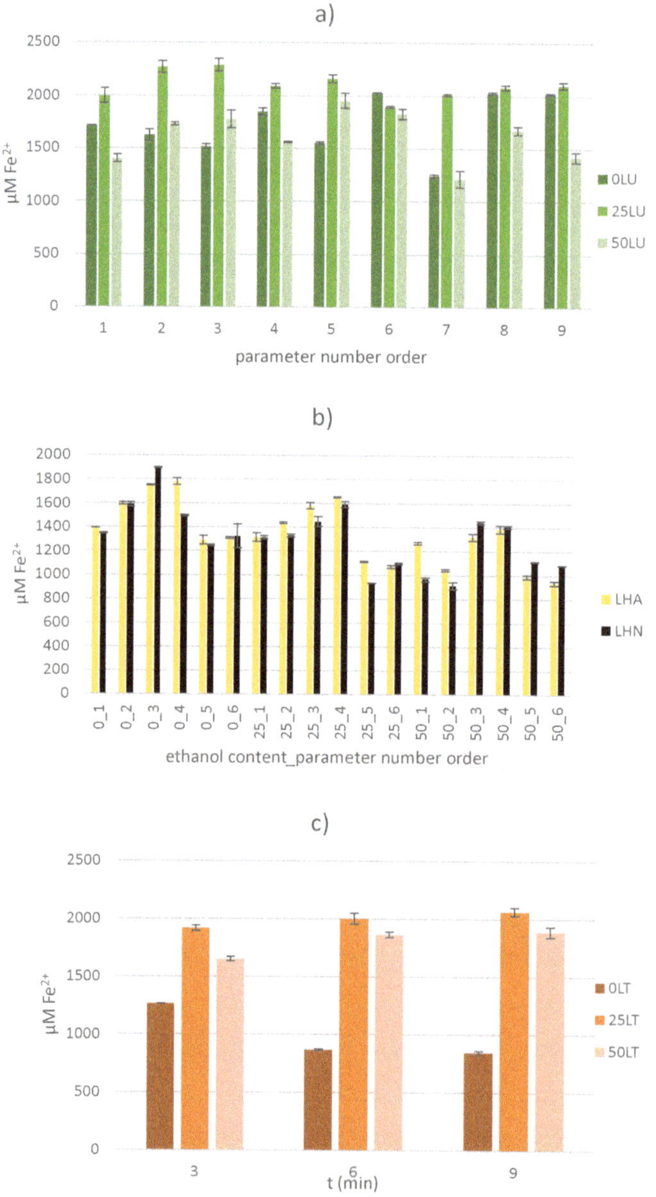

Figure 3. Determination of antioxidant capacity by FRAP method for (**a**) US-treated samples, (**b**) HVED-treated samples and (**c**) thermally treated samples.

Compared to the US-treated samples, lower AC values were found in the thermally treated samples (845.71 ± 16.42 to 2063.55 ± 34.48 μM Fe^{2+}). In contrast to the US-treated samples, the input variable, ethanol content, had a statistically significant influence ($p < 0.05$) on the AC value in the thermally treated samples. A negative influence of treatment time on the ACs of the thermally treated samples when deionized water was used as the extraction solvent was also observed. Specifically, by increasing the treatment time from 3 to 9 min, the AC value decreased 1.50 times. In contrast, the AC value of the thermally treated samples increased when ethanol solutions were used and the treatment time was increased. The use of higher ethanol concentrations does not necessarily guarantee higher AC values. Safdar et al., 2017, reported higher AC values of kinnow peel samples treated with 80% alcohol solutions than with 100% alcohol solutions [44].

The lowest AC values were found in the HVED-treated samples, which ranged from 912.75 ± 28.75 to 1898.20 ± 4.03 μM Fe^{2+}. A statistically significant influence on the AC value of the HVED-treated samples was found for all input variables except the variable of applied gas. Although 61.11% of the LHA samples had a higher AC value, this variable did not have a statistically significant effect on the AC value ($p > 0.05$). The treatment time had a statistically positive influence ($p < 0.05$) on the AC value, while the ethanol content and the applied voltage had a negative influence on the same value ($p < 0.05$). The obtained results do not correlate with the results of studies on HVED-treated samples of thyme leaves [45] and olive leaves [46]. In particular, in the study using thyme leaves, the highest AC value was found when argon gas was applied, and a positive influence of the applied voltage and ethanol content was observed as statistically significant input variables ($p < 0.05$) [45]. Considering the different sample types and their chemical composition, the obtained results are not unexpected (the TPC yield of thyme leaf samples is 6.04 times higher than the TPC yield of sugar beet leaf samples). Namely, the sugar beet leaves are a good source of proteins and bioactive compounds with antioxidant activity that work on the principle of hydrogen transfer and therefore cannot be determined by this method and consequently cannot contribute to the AC results [36,37,47].

2.3.2. DPPH (2,2-Diphenyl-2-picrylhydrazyl) Free Radical Assay

The DPPH method is another spectrophotometric method of determining the antioxidant capacity of samples. Similar to the FRAP method, it is a quick, simple and inexpensive method that does not require any special equipment. Like any in vitro method for the determination of AC, this method is subject to certain limitations. Since the solubility of DPPH• radicals is limited to organic solvents (ethanol/methanol), the results of measuring the ACs of hydrophilic compounds with antioxidant activity are questionable [48]. Furthermore, as with the FRAP method, certain compounds absorb at the same measurement wavelength, which contributes to high false positive AC values. Due to the spatial limitation of the reaction between DPPH• radicals and large molecules with antioxidant activity, the reaction proceeds slowly or not at all, resulting in falsely lower AC values [49]. In addition, it is not suitable for the determination of AC in samples with proteins (which also possess antioxidant activity), as protein precipitation occurs in alcoholic solutions, which also leads to falsely low AC values [50]. Nevertheless, the AC value was determined in all of the samples using the DPPH method and the obtained results are presented graphically (Figure 4).

Regardless of the extraction method used, the AC values determined by the DPPH method were significantly lower than those measured by the FRAP method. In particular, values that were 2.69–5.15 times lower were measured in the US-treated samples, 3.06–7.95 in the thermally treated samples and 2.95–13.40 in the HVED-treated samples. These observations are consistent with studies on US-treated samples of makiang seeds, where the AC values measured by the DPPH method were up to 1.51 times higher than those measured by the FRAP method [51]. In contrast, HVED-treated thyme samples showed higher values (by more than 15 times) of the AC measured by the DPPH method [45]. As already mentioned, the chemical composition of the plant material has a great influence on the final value of the AC and thus on the relationship between the determination methods used. In this work, although the AC values measured by the DPPH method were lower, the trend of increase/decrease between extraction methods and within samples for each extraction method was similar to the trend determined by the FRAP method. In particular, no single statistically significant influence of the input variables was found in the US-treated samples, but a statistically negative influence of the mutual interaction of the amplitude and the ethanol content on the AC value was determined by the DPPH method ($p < 0.05$). The highest AC value (in general, but also within the US-treated samples) of 757.29 ± 1.82 μM TE was observed in the US-treated 25LU9 sample (with 25% ethanol solution as the extraction solvent, an applied amplitude of 50% and a treatment time of 9 min). An increase in the AC value using 25% ethanol solution (extraction solvent) was also observed. The AC values of the US-treated samples ranged from 292.07 ± 7.32 to 757.29 ± 1.82 μM TE and increased 1.50 times compared to the HVED-treated samples and 1.16 times compared to thermally treated samples. The obtained results are not consistent with the research results on olive kernel samples. In fact, in the aforementioned study, higher values of the AC were measured with the HVED extraction method (at a voltage of 40 kV) than with pulsed electric field or ultrasound methods [52]. The application of a significantly higher voltage (15 or 20 kV higher) contributed to the appearance of more expressive electrical charges, which consequently led to a stronger and more expressive fragmentation of the plant material. This had a positive effect on the destruction of complexes between phenolic compounds and other compounds (such as sugars, proteins, etc.) and contributed to increasing the yield of components with antioxidant activity [53].

In the samples of sugar beet leaves treated with HVED, the treatment time proved to be a statistically significant positive input variable for the AC value measured by the DPPH method ($p < 0.05$). Similarly to the FRAP method, the gas used had no statistically significant effect on the value of the AC ($p > 0.05$), with a higher AC value found in 72.22% of the samples in which argon gas was used. The obtained results correlate with the results observed in HVED-treated thyme leaf samples [45]. In addition to the statistically positive influence of the treatment time, the interaction between the treatment time and the gas used had a statistically significant negative influence on the AC value ($p < 0.05$). In general, the reasons for the lower AC value in the HVED-treated samples (compared to the US-treated samples) have already been explained, and the values ranged from 133.00 ± 10.30 to 504.14 ± 1.82 μM TE.

A similar range of values (109.21 ± 4.55 to 655.57 ± 103.06 μM TE) was found for the thermally treated samples. The highest AC values of 655.57 ± 103.06 μM TE were recorded in the TL25/9 sample (with 25% ethanol solution as the extraction solvent and a treatment time of 9 min) and the lowest values of 109.21 ± 4.55 μM TE were recorded in the TL0/6 sample (with deionized water as the extraction solvent and a treatment time of 6 min). The treatment time showed no statistically significant influence on the AC value ($p > 0.05$), but the ethanol content did ($p < 0.05$). As with the determination of AC by the FRAP method, the highest values of AC determined by the DPPH target in thermally treated samples were observed when a 25% ethanol solution was used as the extraction solvent. The recorded results are consistent with the research results on thermally treated (100 °C) samples of white mulberry. Namely, with a longer treatment time (from 15 to 45 min) and the application of a higher ethanol content (from 0% to 70% ethanol solution), the AC value measured by the DPPH method, expressed as radical scavenging (%), increased by 1.24 to 1.59 times [54].

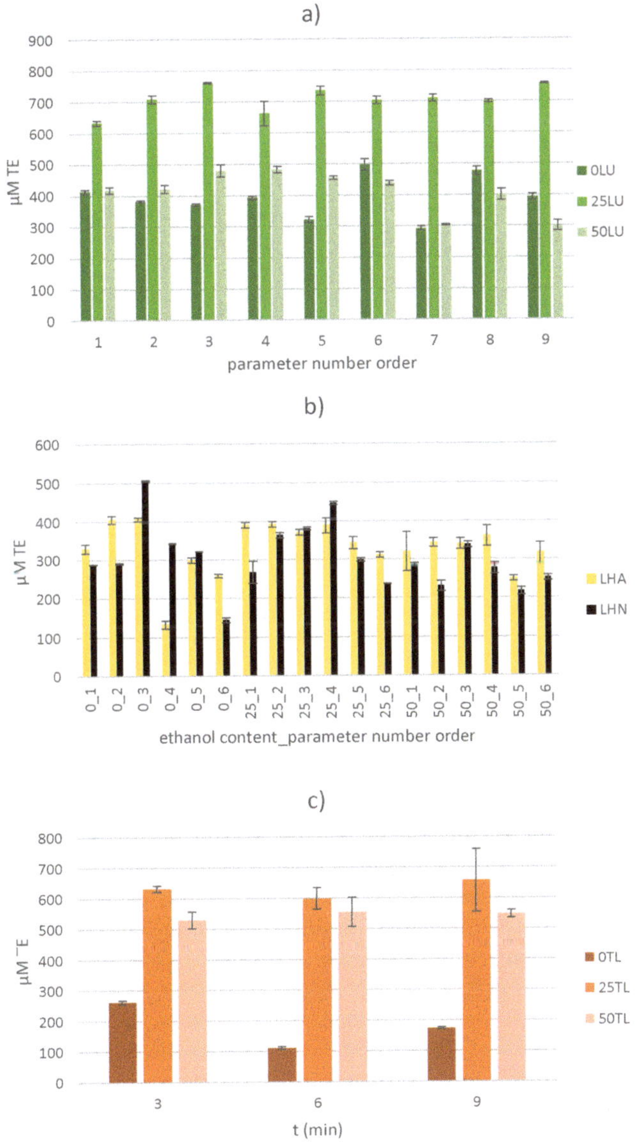

Figure 4. Determination of antioxidant capacity by DPPH method for (**a**) US-treated samples, (**b**) HVED-treated samples and (**c**) thermally treated samples.

2.3.3. Determination of Antioxidant Capacity by Electron Spin Resonance (ESR) Spectroscopy

Figure 5a shows the results of the ESR measurements obtained in US-treated samples (from 118.42 ± 7.17 to 280.93 ± 7.32 µM TE), using various amplitudes, treatment times and ethanol contents. It can be seen that the US-treated samples with deionized water as the extraction solvent (without ethanol) had lower AC values compared to the other US-treated samples, with the exception of the 0LU6 (227.26 ± 5.15 µM TE), 0LU8 (201.53 ± 6.18 µM TE) and 0LU9 (174.71 ± 16.79 μM TE) samples. These samples were US-treated at maximum amplitude (100%) for 9, 6 and 3 min, respectively. In general, all of the input variables showed a positive statistical significance on the value of the antioxidant capacity, while on the other hand, the interaction between the applied amplitude and ethanol content had a statistically significant negative effect on the AC values ($p < 0.05$). In a previous work [55], the influence of the sonication treatment of water on the formation of hydroxyl radicals was investigated by the EPR spectroscopy/spin-trapping method. It was found that the number of radicals increased with the treatment time and with an increasing amplitude. The results in Figure 5a confirm previous findings. Based on these facts, it can be assumed that the mentioned samples contained a larger number of hydroxyl radicals compared to other US-treated samples with deionized water. Although hydroxyl radicals are short-lived, it is possible that they reacted with some of the components extracted from the sample, resulting in the formation of stable radicals. Such stable radicals could contribute to the increase in the AC value. US-treated samples with an ethanol solution (extraction solvent) showed slightly higher AC values than samples extracted with deionized water. There are two possible explanations for this result: first, the use of a water–ethanol mixture may lead to better extraction and consequently to a higher concentration of antioxidant compounds; and second, the addition of small amounts of ethanol to water leads to the appearance of 1-hydroxyethyl free radicals [56]. These radicals could react with DPPH radical and contribute to an increase in the AC value in the US-treated samples. One could expect that the concentration of free radicals formed from ethanol will increase by increasing the ethanol content in the extraction solvent. However, it has been shown that once the optimal ethanol content is reached, a further increase leads to a decrease in the number of carbon radicals [57]. Therefore, a higher ethanol content, which is particularly pronounced in sample 50LU8 (with a 50% ethanol mixture, 100% amplitude and 9 min treatment time), could lead to a lower number of free 1-hydroxyethyl radicals, as in the case of carbon radicals [57].

In the earlier work mentioned above [55], the influence of HVED-treated water on the formation of hydroxyl radicals was also investigated. It was proven that the plasma-induced formation of hydroxyl radicals depends on several factors, and one of the most important is the type of gas used. Therefore, two series of samples were investigated in the presence of nitrogen (from 53.23 ± 2.58 to 199.21 ± 6.55 μM TE) and argon (from 34.25 ± 2.76 to 181.78 ± 3.57 μM TE) in this work. The results are shown in Figure 5b. Statistically, a significant positive effect ($p < 0.05$) of the treatment time on the AC value was observed for all of the HVED-treated samples, regardless of the gas used. Argon reacts with plasma electrons and then reacts with water molecules to form hydroxyl radicals [58]. Nitrogen molecules react in a very similar way [59]. The highest concentration of hydroxyl radicals was found when using argon mixed with water as a working gas, followed by nitrogen [58]. As already indicated, the addition of even small amounts of ethanol to water leads to the occurrence of carbon radicals, and the concentration increased sharply with further increases in ethanol content, indicating that ethanol molecules are predominantly degraded to carbon radicals [57]. These radicals were presumably primary radicals that could react with other radicals, including DPPH, and contribute to an increase in the AC value of the HVED-treated samples. The results also showed a continuous increase in radical concentration with the increase in ethanol content until the value $\chi = 0.37$ was reached, after which the value started to decrease. All of these facts should be considered when analyzing the results. The 0LHN3 sample, which was HVED-treated with deionized water as an extraction solvent for 9 min at a lower voltage (20 kV) in the presence of nitrogen gas, showed the highest AC value (199.21 ± 6.55 μM TE). This is followed by the 0LHA3 sample (181.78 ± 3.57 μM TE) treated under the same conditions in the presence of argon. However, these two samples were an exception compared to other HVED-treated samples with deionized water (extraction solvent). The addition of ethanol in a lower amount (25%) indicates a positive effect when coupled with argon and a higher voltage (25 kV), while this effect was absent under the same conditions with nitrogen. Under these conditions, argon obviously plays the most important role. In contrast, the presence of nitrogen seems to favor a higher ethanol content with the longest treatment time regardless of the applied voltage (except for Sample 50LHA5). From the presented results, each factor can have a different influence on the AC under different conditions. To determine which factor predominates, further measurements would need to be performed.

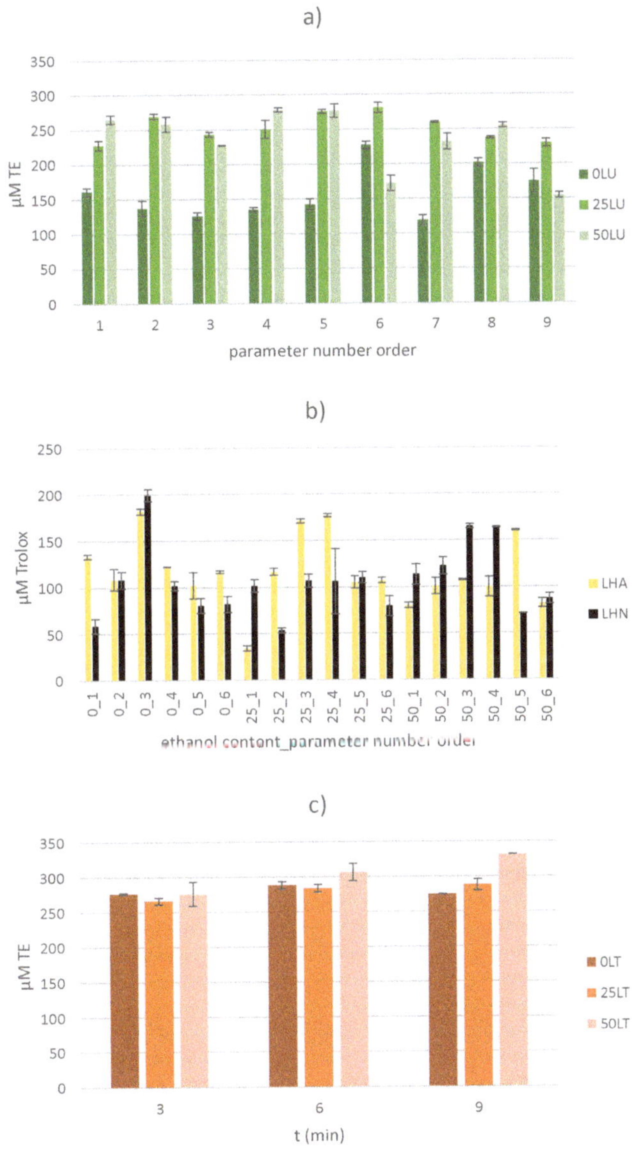

Figure 5. Determination of antioxidant capacity by EPR method for (**a**) US-treated samples, (**b**) HVED-treated samples and (**c**) thermally treated samples.

The effects of thermal treatment on the ESR results are shown in Figure 5c. The temperature influence, regardless of the treatment time (from 3 to 9 min), did not show a noticeable effect on the AC value of the thermally treated samples when deionized water was used as an extraction solvent. However, when all of the thermally treated samples were considered, the treatment time, ethanol content and the interaction between treatment time and the ethanol content showed a positive statistical significance on the AC value ($p < 0.05$). With the addition of the 25% ethanol solution, a slight increase in the AC value can be observed with an increasing treatment time. This effect was more pronounced with an ethanol content of 50%. The maximum AC value (331.36 ± 0.32 μM TE) was reached at the longest treatment time (9 min) with 50% ethanol solution (extraction solvent). This result indicates that ethanol in combination with deionized water improves the extraction of the bioactive components. The positive influence of thermal treatment could be attributed to the increase in the TPC and total flavonoid content (TFC), as shown in the garlic samples [60]. In contrast to the results of that study, this study's results indicate a positive correlation between treatment time and AC ($p < 0.05$). However, it should be noted that the treatment times in this study were shorter, namely 3–9 min, compared to 15–60 min.

2.4. Plasma Characterization by Optical Emission Spectroscopy (OES)

Optical emission spectra (OES) are shown in Figure 6 of (a) argon and (b) nitrogen plasma. In each case, the spectra of sugar beet leaves immersed in water (black line), 25% ethanol (red line) and 50% ethanol (blue line) during plasma treatment are shown. All of the emitted lines were identified using the NIST Atomic Spectra Database [61] or consulting the literature [62–65]. In Figure 6a, the Ar plasma is shown with the dominant emission from neutral atomic argon and oxygen (the emission lines are saturated), while atomic hydrogen from the Balmer series is present with a distinctive emission line at 656 nm, ascribed to Hα. A weak molecular hydrogen emission (Fulcher band) is observed around 600 nm. Qualitatively, there are no significant differences between the spectra for water, 25% ethanol and 50% ethanol. The presence of atomic oxygen and hydrogen in all of the spectra indicates that plasma interacts mainly with the air molecules dissociating hydrogen and oxygen molecules and with the argon atoms emerging from the liquid. Figure 6b shows nitrogen plasma in which the emission of nitrogen species dominates, while atomic hydrogen is also present. It is evident that in the blue part of the spectrum emission, molecular N_2 (2nd positive band) and N_2^+ (1st negative band) are present, while across the entire optical spectrum, atomic and ionic nitrogen is observed. Atomic hydrogen shows a clear emission at 656 nm. Again, no differences are observed between the spectra recorded for deionized water, 25% ethanol and 50% ethanol. In both cases, in argon and nitrogen plasmas, the emission of carbon, as a possible product of ethanol dissociation, was not observed. This indicates that the plasma does not interact with the ethanol in manner that decomposes the structure of the liquid.

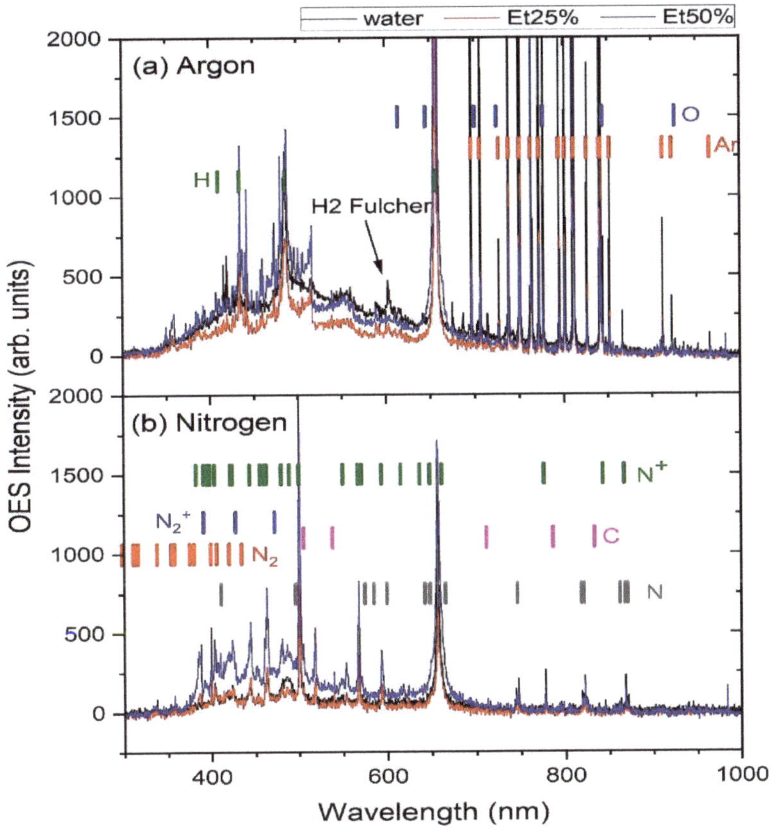

Figure 6. OES spectra of (**a**) argon and (**b**) nitrogen plasmas.

3. Materials and Methods

3.1. Chemicals

All of the chemicals used in this work corresponded to p.a. purity grade or higher. For HPLC-MS/MS analysis, deionized water was purified using a water purification system from Siemens Ultra Clean (Munich, Germany); otherwise, deionized water from the Faculty of Food Technology and Biotechnology, Zagreb, Croatia, was used. Anhydrous sodium carbonate was purchased from K.T.T.T. (Sveta Nedjelja, Croatia). Gallic acid, 6-hydroxy-2,5,7,8-tetramethylchroman-2-carboxylic acid (Trolox) and 2,4,6-Tripyridyl-S-triazine (TPTZ) standards were purchased from Acros Organics (Geel, Belgium). Vitexin standard (4′,5,7-trihydroxyflavone-8-glucoside) and formic acid for mobile phase were purchased from Sigma-Aldrich (St. Louis, MO, USA). Methanol, acetonitrile and iron (II)-sulfate heptahydrate were bought from Honeywell (Seelze, Germany). Iron (III) chloride, sodium acetate trihydrate and ethanol were purchased from GRAM MOL (Zagreb, Croatia). 2,2-Diphenyl-2-Picrylhydrazyl (DPPH) was bought from ABCR (Karlsruhe, Germany) and stored in the dark at 4 °C. Folin-Ciocalteu reagent purchased from Kemika (Zagreb, Croatia) was stored under the same conditions. Hydrochloric acid was bought from Carlos Erba Reagents (Val-de-Reuil, France) and glacial acetic acid was purchased from J.T. Baker (Gliwice, Poland).

3.2. Plant Material

Dry sugar beet (*Beta vulgaris* L.) leaves with a certain dry matter of 94.49 ± 1.6% were provided by project partners from Turkey (Kayseri Şeker, Kocasinan Kayseri, Turkey), the same as in the previous study [47], and in order to facilitate extraction during sample preparation, dry leaves were ground to the particle size distribution of d(0.1) ≤ 238.490 μm; d(0.5) ≤ 630.116 μm; d(0.9) ≤ 1196.769 μm. The particle size was determined using the laser particle size analyzer Mastersizer 2000 (Malvern Instruments GmbH, Herrenberg, Germany).

3.3. Extraction Methods and Sample Labeling

3.3.1. High-Power Ultrasound Assisted Extraction (US)

A total of 27 samples of sugar beet leaves were ultrasonically treated with Q700 Sonicator (Qsonica, Newtown, CT, USA) by adding 100 mL of extraction solvent (deionized water and 25% and 50% ethanol solutions) into a 250 mL double-walled laboratory beaker with 2 ± 0.001 g of weighed ground sugar beet leaves. The position of the ultrasonic probe (d 12 mm) is described in detail by Dukić et al. [19]. To avoid overheating, high-capacity recirculating chiller 4905 (Qsonica, Newtown, CT, USA) was used. The name of the sample consists of an alphanumeric designation, whereby the number before the letter indicates the value of the ethanol content (0, 25 and 50%), the letters indicate the type of treatment (LU, sugar beet leaves US- treated) and the number after the letters indicates the specified US process parameters. Sample 25LU4, for example, represents a sample that was US-treated with 25% ethanol as an extraction solvent and with an amplitude of 50% over a treatment time of 9 min. For a better understanding, the labeling of the samples and the US process parameters are listed in Table 1.

Table 1. Sample labeling and process parameters for US-treated samples.

Ethanol Content (%)	US-Treated	Parameter Number	Amplitude (%)	Treatment Time (min)
0 25 50	LU	**1**	75	6
		2	75	3
		3	50	6
		4	50	9
		5	75	9
		6	100	9
		7	50	3
		8	100	6
		9	100	3

3.3.2. High-Voltage Electrical Discharge-Assisted Extraction (HVED)

In addition to the US method, another non-thermal extraction method was used. To compare non-thermal and thermal methods, high-voltage electrical discharge-assisted extractions (HVED) were performed as well. A total of 1 ± 0.001 g of ground, dry sugar beet leaf samples was mixed with 50 mL of extraction solvent (deionized water and 25% and 50% ethanol solutions) in a 100 mL reactor. A total of 36 samples of sugar beet leaves were HVED-treated using the IMP-SSPG-1200 generator (Impel group d.o.o., Zagreb, Croatia). A frequency of 100 Hz, a pulse width of 400 ns, a distance between the electrodes of 15 mm and a gas flow of 0.75 L/min were previously described in detail by Nutrizio et al. [66]. The applied HVED treatment parameters were a voltage of 20 and 25 kV for nitrogen and argon gas over a treatment time of 3, 6 and 9 min. Same as with the US-treated samples, the sample name consists of an alphanumeric designation, where the number before the letter indicates the ethanol content value (0, 25 and 50%), the letters LH indicate the HVED treatment, the letter A (stands for "argon") or N (stands for "nitrogen") indicate the used gas and the number after the letters indicates the HVED process parameters. For example, Sample 0LHN1 represents an HVED-treated sample with deionized water as extraction solvent and nitrogen as applied gas over a treatment time of 6 min at 25 kV. For a better understanding, the names of the samples and the HVED process parameters are listed in Table 2.

Table 2. Sample labeling and process parameters for HVED-treated samples.

Ethanol Content (%)	HVED-Treated	Applied Gas (A/N)	Parameter Number	Voltage (kV)	Treatment Time (min)
0 25 50	LH	A (Argon) N (Nitrogen)	1	25	6
			2	20	6
			3	20	9
			4	25	9
			5	20	3
			6	25	3

3.3.3. Conventional Thermal Extraction

The thermal extraction of bioactive compounds and antioxidant capacity from sugar beet leaves using the conventional method was performed using a water bath with a mixing function (Memmert GmbH, Schwabach, Germany). The temperature in the water bath was set to 60 °C and stirring was carried out at 150 rpm. A total of 2 ± 0.1000 g of ground dry sugar beet leaves was weighed into a 250 mL laboratory beaker and 100 mL of extraction solvent (deionized water and 25% and 50% ethanol solutions) was added. The samples were placed in the water bath and left in the water bath for 3, 6 and 9 min. The labeling of the thermally treated samples is shown in Table 3.

Table 3. Sample labeling and process parameters for thermally treated samples.

Sample Name	Ethanol Content (%)	Treatment Time (min)
0LT3	0	3
0LT6	0	6
0LT9	0	9
25LT3	25	3
25LT6	3	6
25LT9	25	9
50LT3	50	3
50LT6	50	6
50LT9	50	9

3.4. Analysis

3.4.1. Determination of Total Phenolic Content (TPC)

The total concentration of phenolic compounds in the samples was measured with the UV-VIS spectrophotometer UV-2600i (Shimadzu, Kyoto, Japan) at a wavelength of 765 nm and according to the Folin–Ciocalteu method [67], using gallic acid (GAE) as a standard and calculated using the following linear equation:

$$y = 0.00107826x + 0.111618 \quad (1)$$

where "y" represents the measured absorbance (at 765 nm), and "x" the equivalent GAE concentration (in mg/L). The sample preparation and measurement procedure are described in detail in a previous article [19]. Taking into account the known mass of the sugar beet leaves and the percentage of dry matter (94.49 ± 1.6%), the results were expressed in $mg/g_{d.m}$.

3.4.2. Determination of Vitexin Content

Standard Solutions and Samples for HPLC-MS/MS Determination

The corresponding amounts of the reference substances were dissolved in methanol to obtain the stock solutions for vitexin. The working solutions used for further investigations were prepared by diluting the stock solutions with methanol or the mobile phase. All of the solutions and samples prepared for high-performance liquid chromatography coupled with mass spectrometry (HPLC-MS/MS) analysis were filtered through a 0.20 μm filter syringe (Filtres Fioroni, Senigallia, Italy) before use. All of the samples were injected in duplicate.

Chromatographic and Mass Spectrometric Conditions

The chromatographic analysis was carried out using an Eksion LC HPLC system (SCIEX, Framingham, MA, USA). The analytes were separated on a Phenomenex column Luna Omega 3 μm Polar C18 100 Å, 100 × 4.6 mm (Torrance, CA, USA), with thermostat column temperature of 40 °C, automated sampling temperature of 4 °C and injection volume of 10 μL. The mobile phases consisted of: A 100% H_2O with 0.01% HCOOH (*v*/*v*) and B 100% ACN with 0.01% HCOOH (*v*/*v*) at a flow rate of 0.40 mL/min. The gradient was set as follows: 1 min 10% B, 13 min 90% B, 15 min 90% B, 16 min 10% B. Column efflux was monitored using a SCIEX 4500 QTRAP (SCIEX, Framingham, MA, USA) triple quadrupole mass spectrometer with an electrospray ionization (ESI) source. Vitexin was ionized in negative electrospray ionization mode, with ionization temperature of 300 °C, ion spray voltage of −4500 V, drying gas temperature of 190 °C and drying gas flow of 9.0 L/min [68]. Nitrogen was used as nebulizing, curtain and collision gas. The mass spectrometer was programmed for multiple reaction monitoring (MRM). The precursor ion [M − H]—at *m*/*z* 431.0 was monitored via the first quadrupole filter (Q1), while the product ion at *m*/*z* 311.1, 283.1 and 117.0 was monitored via the third quadrupole filter (Q3) [69]. Declustering potential (DP) was −120 V, the collision energy (CE) was −32.0 and the collision cell exit potential (CXP) was −11.0. The peak areas obtained from MRM were used for the quantification of vitexin [70]. The data were processed with Multiquant 3.6 software (SCIEX, Darmstadt, Germany).

Method Validation

For the chromatographic determination's linearity test, standard vitexin solutions were injected in triplicate in seven steps over a concentration range of 0.10 to 100 ng/mL. Good linearity was obtained, with correlation coefficients between 0.99956 and 0.9987. The limit of detection (LOD) and the limit of quantification (LOQ) were investigated by stepwise dilution of each standard solution with methanol; LOD and LOQ were calculated for a ratio S/N of 3 and 10, respectively. The LOQ for chromatographic determination of vitexin quantification was confirmed by injecting the 6-fold vitexin standard with a concentration of 0.50 ng/mL, i.e., the second point of the calibration curve. The recovery was between 90.34% and 110.63%; the average recovery was 105.70%. The standard deviation (SD) was 0.0377 ng/mL. The repeatability and reproducibility of the method was expressed as relative standard deviation (RSD) and amounted to 7.44%.

The determination of vitexin content in the samples was calculated using the following linear equation:

$$y = 28480.20214x \quad (2)$$

where "y" represents the area under the peak and "x" the equivalent vitexin concentration (in ng/mL).

3.4.3. Determination of Antioxidant Capacity (AC)

Ferric Reducing Antioxidant Power (FRAP) Assay

This method is based on the reduction reaction of the yellow-colored complex iron-2,4,6-tripyridyl-s-triazine (TPTZ) in the presence of antioxidants, resulting in a blue-colored product. The reaction takes place in an acidic medium at pH = 3.6 to ensure good solubility of the iron. At lower pH values, the ionization potential, which enables electron transfer, decreases, and at the same time the redox potential increases, which additionally promotes the reaction in the direction of electron transfer. The redox potential of the Fe(III)/Fe(II) reaction is 0.77 V, and all compounds with a lower redox potential participate in the iron reduction reaction and thus contribute to the final result of AC [71].

The AC in the samples was measured with the UV-VIS spectrophotometer UV-2600i (Shimadzu, Kyoto, Japan) at a wavelength of 595 nm and according to the FRAP method [46] using $FeSO_4 \cdot 7H_2O$ as a standard and calculated using the following linear equation:

$$y = 6.81142 * 10^{-4}x + 0.00843793 \quad (3)$$

where "y" represents the measured absorbance (at 595 nm) and "x" the equivalent Fe^{2+} concentration (in μM). The chemicals/sample preparation and measurement procedure are described in detail in a previous article [46].

DPPH (2,2-Diphenyl-2-picrylhydrazyl) Free Radical Assay

This spectrophotometric method for the determination of AC is based on the reduction of the 2,2-diphenyl-2-picrylhydrazyl (DPPH) radical in a methanol solution. The DPPH radical shows strong absorption in the visible region of the spectrum (515 nm) due to the unpaired electron. In the presence of an electron donor, an antioxidant that "quenches" free radicals, DPPH radicals are stabilized by electron pairing, resulting in a color change of the solution from purple (oxidized form of the radical) to yellow (reduced form of the radical). AC is measured by the change in absorbance of the reaction mixture at a given time compared to the same effect obtained with a known reducing agent [72].

The AC in the samples was measured with the UV-VIS spectrophotometer UV-2600i (Shimadzu, Kyoto, Japan) at a wavelength of 515 nm and according to the modified DPPH method [73] using Trolox as a standard and calculated using the following linear equation:

$$y = 0.0007x + 0.0032 \quad (4)$$

where "y" represents the measured absorbance (at 515 nm) and "x" the equivalent Trolox (TE) concentration (in μM). To determine the AC, 100 μL of the sample was pipetted into a test tube and 3.9 mL of a 0.094 mM DPPH solution was added and shaken. The reaction took place in the dark for 30 min (time to reach equilibrium). The measured absorbance was compared with the absorbance of the blank sample (which contained methanol instead of the sample).

Determination of Antioxidant Capacity by Electron Spin Resonance (ESR) Spectroscopy

ESR spectroscopy is a technique that enables the direct detection of free radicals in the sample and is therefore very suitable for measuring the antioxidant capacity of different types of samples [74,75]. The ESR measurements have been performed at Ruđer Bošković Institute in Zagreb, Croatia, at room temperature on a Bruker EMX CW spectrometer (Bruker, Billerica, MA, USA) equipped with the Varian V-7300 magnet (Palo Alto, CA, USA) and Bruker microwave bridge ER 041 XG. The spectroscopic parameters were microwave frequency of 9.3 GHz, power attenuation of 10 mW, magnetic field modulation amplitude of 1 G and modulation frequency of 100 kHz. Spectra were recorded around the central field of 3315 G, with a sweep range of 100 G and sweep time of 20.48 s, 20 min after bringing the DPPH solution into contact with sample extracts. Spectral analysis was carried out with spectrometer build-in analysis WinEPR software (https://www.bruker.com/en/products-and-solutions/mr/epr-instruments/epr-software/winepr.html, accessed on 5 February 2025, Bruker, Billerica, MA, USA) by measuring the intensity of the central line (peak-to-peak amplitude), which is proportional to the number of spins in the sample. All of the measurements were performed in duplicate and in an analog fashion; the samples were put into the capillary which was then placed in ESR tube. The tube was immediately put in Bruker Cylindrical Super-High Q Resonator ER 4122 SHQE and measured using the given parameters. ESR results are presented in Trolox equivalents. Trolox solutions of different concentrations and samples for ESR measurements were prepared in the same way as that used for spectrophotometric determination of AC (DPPH Free Radical Assay). The AC in the samples was determined using Trolox as a standard and calculated using the following linear equation:

$$y = 56.865x + 10.164 \tag{5}$$

where "y" represents the ESR signal (ARB.U) and "x" the equivalent Trolox concentration (in μM).

3.4.4. Plasma Characterization by Optical Emission Spectroscopy (OES)

The experimental setup is shown in Figure 7. It consists of two parts: a reaction chamber and a spectrometer for optical emission spectroscopy (OES) measurement performed during the treatment of the samples. The reaction chamber is made of borosilicate glass and enables the monitoring of the emission in the optically visible part of the emission spectrum (from 350 nm onwards). The chamber consists of two electrodes; the upper one is grounded; the lower one has an opening for the gas inlet (argon or nitrogen) and is immersed in water. The latter is connected to a bias voltage of 20 kV at 100 Hz (Impel group d.o.o., Zagreb, Croatia). The gas flow is controlled and optimized with a flow meter (0.75 L/min).

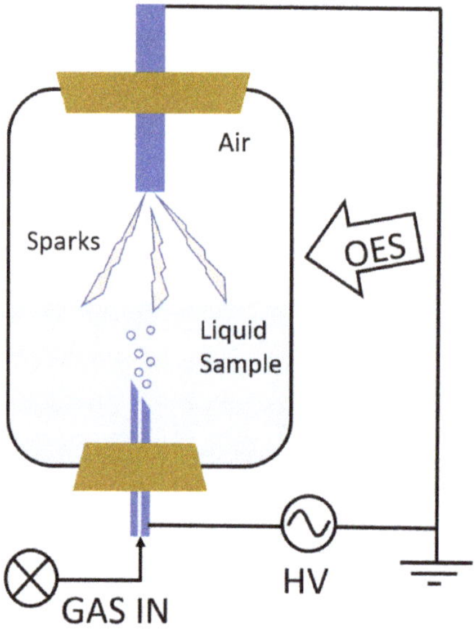

Figure 7. Experimental setup for plasma characterization.

OES of Ar and N_2 plasmas was performed using the HR2000CG-UV-NIR spectrometer (Ocean Optics, Duiven, The Netherlands), which has a spectral resolution of 1 nm and a spectral range of 200–1100 nm. The signal was collected using a fiber (solar resistant, with 600 μm diameter and with a 5 cm lens at the end to gather more light) placed above the water surface. It thus captures all of the light emitted by the discharge sparks interacting with the water surface and the air above it. As a rule, the spectra were recorded with an integration time of 1 s and without averaging. All of the collected spectra were not corrected to a spectral response of the spectrometer, as they were only used to identify the emitted species and not for detailed findings.

3.4.5. Experimental Design and Statistical Analysis

Experimental design and statistical analysis for all of the extraction parameters was performed using STATGRAPHICS Centurion (StatPoint Technologies, Inc., Warrenton, VA, USA). The experimental design included 27 US-treated samples, 36 HVED-treated samples and 9 thermally treated samples. A multilevel factorial design was used to determine the effects of the independent variables on the dependent variables. The extraction parameters had a statistically significant effect if $p < 0.05$, meaning that they were significantly different from zero at the 95.0% confidence interval.

4. Conclusions

The sugar beet leaf, a waste product of the sugar industry, proved to be an excellent raw material for the extraction of bioactive compounds. The US method, as one of the non-thermal extraction methods, proved to be the most suitable method for extracting the bioactive compounds and obtaining the highest AC values from the sugar beet leaves. Although in vitro methods have been successfully used to determine the AC, in vivo methods need to be performed in future research to obtain real values. Indeed, sugar beet leaves are a good source of bioactive proteins such as ribulose-1,5-bisphosphate carboxylase oxygenase enzyme (RuBisCO), whose AC was most likely not detected by spectrophotometric methods due to the nature of the compound itself (protein) and the aforementioned limitations of the DPPH and FRAP methods. For a better interpretation of the results, it is also necessary to analyze the bioavailability of the extracted bioactive compounds. Furthermore, in addition to changing the determination approach, the spectrum of the green solvents used (e.g., natural deep eutectic solvents—NADES) could also be tested in future research. In any case, the results are promising, and it is to be expected that these food wastes will increasingly be analyzed using non-thermal extraction methods.

Author Contributions: Conceptualization, A.R.J., J.D., I.D., S.V. and N.K.; methodology, J.D., I.D., J.J., R.R. and A.R.J.; software, J.D., I.D., J.J. and N.K.; validation, A.R.J., I.D., S.V., D.M. and N.K.; formal analysis, J.D., I.D., J.J. and R.R.; investigation, J.D., I.D., J.J. and N.K.; resources, A.R.J. data curation, J.D. and I.D.; writing—original draft preparation, J.D., I.D., S.V. and N.K.; writing—review and editing, J.D., I.D., S.V., N.K. and M.N.; visualization, J.D. and M.N.; supervision, A.R.J., I.D., S.V., N.K.; project administration, A.R.J.; funding acquisition, A.R.J. and J.D. All authors have read and agreed to the published version of the manuscript.

Funding: This research was funded by the Partnership for Research and Innovation in the Mediterranean Area PRIMA H2020 GA2032: "FunTomP—Functionalized Tomato Products". The work of doctoral student Josipa Dukić has been fully supported by the "Young researchers' career development project—training of doctoral students" of the Croatian Science Foundation (DOK-2021-02). Furthermore, this work was partly supported by the Croatian Science Foundation (HRZZ) project IP-2022-10-9292 and partly supported by the project Centre for Advanced Laser Techniques (CALT), co-funded by the European Union through the European Regional Development Fund under the Competitiveness and Cohesion Operational Program (KK.01.1.1.05.0001).

Data Availability Statement: Data is contained within the article.

Acknowledgments: The authors would like to thank the Partnership for Research and Innovation in the Mediterranean Area PRIMA H2020 GA2032: "FunTomP—Functionalized Tomato Products".

Conflicts of Interest: The authors declare no conflicts of interest.

References

1. Adam, D. How Far Will Global Population Rise? Researchers Can't Agree. *Nature* **2021**, *597*, 462–465. [CrossRef]
2. van Dijk, M.; Morley, T.; Rau, M.L.; Saghai, Y. A Meta-Analysis of Projected Global Food Demand and Population at Risk of Hunger for the Period 2010–2050. *Nat. Food* **2021**, *2*, 494–501. [CrossRef] [PubMed]
3. Arshad, R.N.; Abdul-Malek, Z.; Roobab, U.; Munir, M.A.; Naderipour, A.; Qureshi, M.I.; El-Din Bekhit, A.; Liu, Z.W.; Aadil, R.M. Pulsed Electric Field: A Potential Alternative towards a Sustainable Food Processing. *Trends Food Sci. Technol.* **2021**, *111*, 43–54. [CrossRef]
4. Herrero, M.; Thornton, P.K.; Mason-D'Croz, D.; Palmer, J.; Bodirsky, B.L.; Pradhan, P.; Barrett, C.B.; Benton, T.G.; Hall, A.; Pikaar, I.; et al. Articulating the Effect of Food Systems Innovation on the Sustainable Development Goals. *Lancet Planet Health* **2021**, *5*, e50–e62. [CrossRef]
5. Chemat, F.; Abert Vian, M.; Fabiano-Tixier, A.S.; Nutrizio, M.; Režek Jambrak, A.; Munekata, P.E.S.; Lorenzo, J.M.; Barba, F.J.; Binello, A.; Cravotto, G. A Review of Sustainable and Intensified Techniques for Extraction of Food and Natural Products. *Green Chem.* **2020**, *22*, 2325–2353. [CrossRef]
6. Ng, H.S.; Kee, P.E.; Yim, H.S.; Chen, P.T.; Wei, Y.H.; Chi-Wei Lan, J. Recent Advances on the Sustainable Approaches for Conversion and Reutilization of Food Wastes to Valuable Bioproducts. *Bioresour. Technol.* **2020**, *302*, 122889. [CrossRef] [PubMed]
7. Gallo, M.; Ferrara, L.; Naviglio, D. Application of Ultrasound in Food Science and Technology: A Perspective. *Foods* **2018**, *7*, 164. [CrossRef] [PubMed]
8. Wen, C.; Zhang, J.; Zhang, H.; Dzah, C.S.; Zandile, M.; Duan, Y.; Ma, H.; Luo, X. Advances in Ultrasound Assisted Extraction of Bioactive Compounds from Cash Crops—A Review. *Ultrason. Sonochem.* **2018**, *48*, 538–549. [CrossRef] [PubMed]
9. Lavilla, I.; Bendicho, C. Fundamentals of Ultrasound-Assisted Extraction. In *Water Extraction of Bioactive Compounds: From Plants to Drug Development*; Elsevier: Amsterdam, The Netherlands, 2017; pp. 291–316. [CrossRef]
10. Burm, K.T.A.L. Plasma: The Fourth State of Matter. *Plasma Chem. Plasma Process.* **2012**, *32*, 401–407. [CrossRef]
11. Zahoranová, A.; Hoppanová, L.; Šimončicová, J.; Tučeková, Z.; Medvecká, V.; Hudecová, D.; Kaliňáková, B.; Kováčik, D.; Černák, M. Effect of Cold Atmospheric Pressure Plasma on Maize Seeds: Enhancement of Seedlings Growth and Surface Microorganisms Inactivation. *Plasma Chem. Plasma Process.* **2018**, *38*, 969–988. [CrossRef]
12. Hori, M. Radical-Controlled Plasma Processes. *Rev. Mod. Plasma Phys.* **2022**, *6*, 36. [CrossRef]
13. Shao, T.; Wang, R.; Zhang, C.; Yan, P. Atmospheric-Pressure Pulsed Discharges and Plasmas: Mechanism, Characteristics and Applications. *High Volt.* **2018**, *3*, 14–20. [CrossRef]
14. Zeb, A. Concept, Mechanism, and Applications of Phenolic Antioxidants in Foods. *J. Food Biochem.* **2020**, *44*, e13394. [CrossRef]
15. Wang, T.-y; Li, Q.; Bi, K.-s. Bioactive Flavonoids in Medicinal Plants: Structure, Activity and Biological Fate. *Asian J. Pharm. Sci.* **2018**, *13*, 12–23. [CrossRef] [PubMed]
16. Flieger, J.; Flieger, W.; Baj, J.; Maciejewski, R. Antioxidants: Classification, Natural Sources, Activity/Capacity Measurements, and Usefulness for the Synthesis of Nanoparticles. *Materials* **2021**, *14*, 4135. [CrossRef] [PubMed]
17. Pisoschi, A.M.; Pop, A.; Iordache, F.; Stanca, L.; Predoi, G.; Serban, A.I. Oxidative Stress Mitigation by Antioxidants—An Overview on Their Chemistry and Influences on Health Status. *Eur. J. Med. Chem.* **2021**, *209*, 112891. [CrossRef] [PubMed]
18. Ebrahimi, P.; Mihaylova, D.; Marangon, C.M.; Grigoletto, L.; Lante, A. Impact of Sample Pretreatment and Extraction Methods on the Bioactive Compounds of Sugar Beet (*Beta vulgaris* L.) Leaves. *Molecules* **2022**, *27*, 8110. [CrossRef] [PubMed]
19. Dukić, J.; Hunić, M.; Nutrizio, M.; Režek Jambrak, A. Influence of High-Power Ultrasound on Yield of Proteins and Specialized Plant Metabolites from Sugar Beet Leaves (*Beta vulgaris* subsp. *vulgaris* var. *altissima*) . *Appl. Sci.* **2022**, *12*, 8949. [CrossRef]
20. Tenorio, A.T. Sugar Beet Leaves for Functional Ingredients. Ph.D. Thesis, Wageningen University & Research, Wageningen, The Netherlands, 2017. [CrossRef]
21. Zewail, R.M.Y.; El-Gmal, I.S.; Khaitov, B.; El-Desouky, H.S.A. Micronutrients through Foliar Application Enhance Growth, Yield and Quality of Sugar Beet (*Beta vulgaris* L.). *J. Plant Nutr.* **2020**, *43*, 2275–2285. [CrossRef]
22. El-gengaihi, S.E.; Hamed, M.A.; Aboubaker, D.H.; Mossa, A.H. Flavonoids from sugar beet leaves as hepatoprotective agent. *Int. J. Pharm. Sci.* **2016**, *8*, 281–286.
23. Maravić, N.; Teslić, N.; Nikolić, D.; Dimić, I.; Šereš, Z.; Pavlić, B. From Agricultural Waste to Antioxidant-Rich Extracts: Green Techniques in Extraction of Polyphenols from Sugar Beet Leaves. *Sustain. Chem. Pharm.* **2022**, *28*, 100728. [CrossRef]

24. Drăghici-Popa, A.M.; Boscornea, A.C.; Brezoiu, A.M.; Tomas, Ș.T.; Pârvulescu, O.C.; Stan, R. Effects of Extraction Process Factors on the Composition and Antioxidant Activity of Blackthorn (*Prunus spinosa* L.) Fruit Extracts. *Antioxidants* **2023**, *12*, 1897. [CrossRef]
25. Ebrahimi, P.; Bayram, I.; Mihaylova, D.; Lante, A. A Strategy to Minimize the Chlorophyll Content in the Phenolic Extract of Sugar Beet Leaves: Can This Extract Work as a Natural Antioxidant in Vegetable Oils? *Food Bioprocess. Tech.* **2024**, *18*, 2493–2506. [CrossRef]
26. Hemmati, V.; Garavand, F.; Khorshidian, N.; Cacciotti, I.; Goudarzi, M.; Chaichi, M.; Tiwari, B.K. Impact of Cold Atmospheric Plasma on Microbial Safety, Total Phenolic and Flavonoid Contents, Antioxidant Activity, Volatile Compounds, Surface Morphology, and Sensory Quality of Green Tea Powder. *Food Biosci.* **2021**, *44*, 101348. [CrossRef]
27. Spigno, G.; Tramelli, L.; De Faveri, D.M. Effects of Extraction Time, Temperature and Solvent on Concentration and Antioxidant Activity of Grape Marc Phenolics. *J. Food Eng.* **2007**, *81*, 200–208. [CrossRef]
28. Snoussi, A.; Bouacida, S.; Mitić, M.; Arsić, B.; Koubaier, H.B.H.; Chouaibi, M.; Janković, S.; Zlatanović, I.; Mrmošanin, J.; Stojanović, G.; et al. Thermal Degradation Kinetics of Myrtle Leaves Ethanol Extract (*Myrtus communis* L.): Effect on Phenolic Compounds Content and Antioxidant Activity. *J. Food Meas. Charact.* **2022**, *16*, 2119–2130. [CrossRef]
29. Zu, Y.; Zhang, Q.; Zhao, X.; Wang, D.; Li, W.; Sui, X.; Zhang, Y.; Jiang, S.; Wang, Q.; Gu, C. Preparation and Characterization of Vitexin Powder Micronized by a Supercritical Antisolvent (SAS) Process. *Powder Technol.* **2012**, *228*, 47–55. [CrossRef]
30. Razak, L.T. Optimization of Vitexin and Isovitexin Compounds Extracted from Dried Mas Cotek Leaves Using One-Factor-at-a-Time (OFAT) Approach in Aqueous Extraction. *Int. Food Res. J.* **2018**, *25*, 2560–2571.
31. Sarangapani, C.; O'Toole, G.; Cullen, P.J.; Bourke, P. Atmospheric Cold Plasma Dissipation Efficiency of Agrochemicals on Blueberries. *Innov. Food Sci. Emerg. Technol.* **2017**, *44*, 235–241. [CrossRef]
32. Chen, F.; Zhang, Q.; Liu, J.; Gu, H.; Yang, L. An Efficient Approach for the Extraction of Orientin and Vitexin from Trollius Chinensis Flowers Using Ultrasonic Circulating Technique. *Ultrason. Sonochem.* **2017**, *37*, 267–278. [CrossRef]
33. Zu, G.; Zhang, R.; Yang, L.; Ma, C.; Zu, Y.; Wang, W.; Zhao, C. Ultrasound-Assisted Extraction of Carnosic Acid and Rosmarinic Acid Using Ionic Liquid Solution from *Rosmarinus officinalis*. *Int. J. Mol. Sci.* **2012**, *13*, 11027–11043. [CrossRef]
34. Tiwari, B.K. Ultrasound: A Clean, Green Extraction Technology. *TrAC Trends Anal. Chem.* **2015**, *71*, 100–109. [CrossRef]
35. Chemat, F.; Rombaut, N.; Sicaire, A.G.; Meullemiestre, A.; Fabiano-Tixier, A.S.; Abert-Vian, M. Ultrasound Assisted Extraction of Food and Natural Products. Mechanisms, Techniques, Combinations, Protocols and Applications. A Review. *Ultrason. Sonochem.* **2017**, *34*, 540–560. [CrossRef]
36. Malta, L.G.; Liu, R.H. Analyses of Total Phenolics, Total Flavonoids, and Total Antioxidant Activities in Foods and Dietary Supplements. In *Encyclopedia of Agriculture and Food Systems*; Academic Press: Cambridge, MA, USA, 2014; pp. 305–314. [CrossRef]
37. Gulcin, İ. Antioxidants and Antioxidant Methods: An Updated Overview. *Arch. Toxicol.* **2020**, *94*, 651–715. [CrossRef] [PubMed]
38. Somogyi, A.; Rosta, K.; Pusztai, P.; Tulassay, Z.; Nagy, G. Antioxidant Measurements. *Physiol. Meas.* **2007**, *28*, R41. [CrossRef]
39. Nilsson, J.; Pillai, D.; Önning, G.; Persson, C.; Nilsson, Å.; Åkesson, B. Comparison of the 2,2′-Azinobis-3-Ethylbenzotiazo-Line-6-Sulfonic Acid (ABTS) and Ferric Reducing Anti-Oxidant Power (FRAP) Methods to Asses the Total Antioxidant Capacity in Extracts of Fruit and Vegetables. *Mol. Nutr. Food Res.* **2005**, *49*, 239–246. [CrossRef] [PubMed]
40. Benzie, I.F.F.; Devaki, M. The Ferric Reducing/Antioxidant Power (FRAP) Assay for Non Enzymatic Antioxidant Capacity: Concepts, Procedures, Limitations and Applications. In *Measurement of Antioxidant Activity and Capacity: Recent Trends and Applications*; John Wiley & Sons Ltd.: New York, NY, USA, 2017; pp. 77–106. [CrossRef]
41. Pulido, R.; Bravo, L.; Saura-Calixto, F. Antioxidant Activity of Dietary Polyphenols As Determined by a Modified Ferric Reducing/Antioxidant Power Assay. *J. Agric. Food Chem.* **2000**, *48*, 3396–3402. [CrossRef] [PubMed]
42. Ashokkumar, M. Applications of Ultrasound in Food and Bioprocessing. *Ultrason. Sonochem.* **2015**, *25*, 17–23. [CrossRef] [PubMed]
43. Deng, J.; Xu, Z.; Xiang, C.; Liu, J.; Zhou, L.; Li, T.; Yang, Z.; Ding, C. Comparative Evaluation of Maceration and Ultrasonic-Assisted Extraction of Phenolic Compounds from Fresh Olives. *Ultrason. Sonochem.* **2017**, *37*, 328–334. [CrossRef] [PubMed]
44. Safdar, M.N.; Kausar, T.; Jabbar, S.; Mumtaz, A.; Ahad, K.; Saddozai, A.A. Extraction and Quantification of Polyphenols from Kinnow (*Citrus reticulate* L.) Peel Using Ultrasound and Maceration Techniques. *J. Food Drug Anal.* **2017**, *25*, 488–500. [CrossRef] [PubMed]
45. Nutrizio, M.; Pataro, G.; Carullo, D.; Carpentieri, S.; Mazza, L.; Ferrari, G.; Chemat, F.; Banović, M.; Jambrak, A.R. High Voltage Electrical Discharges as an Alternative Extraction Process of Phenolic and Volatile Compounds from Wild Thyme (*Thymus serpyllum* L.): In Silico and Experimental Approaches for Solubility Assessment. *Molecules* **2020**, *25*, 4131. [CrossRef] [PubMed]
46. Žuntar, I.; Putnik, P.; Kovačević, D.B.; Nutrizio, M.; Šupljika, F.; Poljanec, A.; Dubrović, I.; Barba, F.J.; Jambrak, A.R. Phenolic and Antioxidant Analysis of Olive Leaves Extracts (*Olea europaea* L.) Obtained by High Voltage Electrical Discharges (HVED). *Foods* **2019**, *8*, 248. [CrossRef]

47. Dukić, J.; Košpić, K.; Kelava, V.; Mavrić, R.; Nutrizio, M.; Balen, B.; Butorac, A.; Halil Öztop, M.; Režek Jambrak, A. Alternative Methods for RuBisCO Extraction from Sugar Beet Waste: A Comparative Approach of Ultrasound and High Voltage Electrical Discharge. *Ultrason. Sonochem.* **2023**, *99*, 106535. [CrossRef] [PubMed]
48. Arnao, M.B. Some Methodological Problems in the Determination of Antioxidant Activity Using Chromogen Radicals: A Practical Case. *Trends Food Sci. Technol.* **2000**, *11*, 419–421. [CrossRef]
49. Huang, D.; Boxin, O.U.; Prior, R.L. The Chemistry behind Antioxidant Capacity Assays. *J. Agric. Food Chem.* **2005**, *53*, 1841–1856. [CrossRef]
50. Bedlovičová, Z.; Strapáč, I.; Baláž, M.; Salayová, A. A Brief Overview on Antioxidant Activity Determination of Silver Nanoparticles. *Molecules* **2020**, *25*, 3191. [CrossRef]
51. Sirichan, T.; Kijpatanasilp, I.; Asadatorn, N.; Assatarakul, K. Optimization of Ultrasound Extraction of Functional Compound from Makiang Seed by Response Surface Methodology and Antimicrobial Activity of Optimized Extract with Its Application in Orange Juice. *Ultrason. Sonochem.* **2022**, *83*, 105916. [CrossRef]
52. Roselló-Soto, E.; Barba, F.J.; Parniakov, O.; Galanakis, C.M.; Lebovka, N.; Grimi, N.; Vorobiev, E. High Voltage Electrical Discharges, Pulsed Electric Field, and Ultrasound Assisted Extraction of Protein and Phenolic Compounds from Olive Kernel. *Food Bioproc. Technol.* **2015**, *8*, 885–894. [CrossRef]
53. Boussetta, N.; Vorobiev, E. Extraction of Valuable Biocompounds Assisted by High Voltage Electrical Discharges: A Review. *Comptes Rendus Chim.* **2014**, *17*, 197–203. [CrossRef]
54. Cui, W.S.; Zhao, X.H. Composition and Activity Changes of the Soluble Water and Ethanol Extracts from White Mulberry (*Morus alba* L.) Fruits in Response to Thermal Treatment. *J. Food Meas. Charact.* **2020**, *14*, 838–848. [CrossRef]
55. Režek Jambrak, A.; Ojha, S.; Šeremet, D.; Nutrizio, M.; Maltar-Strmečki, N.; Valić, S.; Gajdoš Kljusurić, J.; Tiwari, B. Free Radical Detection in Water after Processing by Means of High Voltage Electrical Discharges and High Power Ultrasound. *J. Food Process. Preserv.* **2021**, *45*, e15176. [CrossRef]
56. Zhang, Q.A.; Shen, Y.; Fan, X.H.; García Martín, J.F.; Wang, X.; Song, Y. Free Radical Generation Induced by Ultrasound in Red Wine and Model Wine: An EPR Spin-Trapping Study. *Ultrason. Sonochem.* **2015**, *27*, 96–101. [CrossRef] [PubMed]
57. Xue, Z.D.; Zhang, Q.A.; Zheng, H.R. Roles of Free Radical on the Formation of Acetaldehyde in Model Wine Solutions under Different Ultrasound Parameters: A Key Bridge-Link Compound for Red Wine Coloration during Ageing. *Ultrason. Sonochem.* **2021**, *79*, 105757. [CrossRef] [PubMed]
58. Takamatsu, T.; Uehara, K.; Sasaki, Y.; Miyahara, H.; Matsumura, Y.; Iwasawa, A.; Ito, N.; Azuma, T.; Kohno, M.; Okino, A. Investigation of Reactive Species Using Various Gas Plasmas. *RSC Adv.* **2014**, *4*, 39901–39905. [CrossRef]
59. Uhm, H.S. Generation of Various Radicals in Nitrogen Plasma and Their Behavior in Media. *Phys. Plasmas* **2015**, *22*, 123506. [CrossRef]
60. Alide, T.; Wangila, P.; Kiprop, A. Effect of Cooking Temperature and Time on Total Phenolic Content, Total Flavonoid Content and Total in Vitro Antioxidant Activity of Garlic. *BMC Res. Notes* **2020**, *13*, 564. [CrossRef] [PubMed]
61. Atomic Spectra Database | NIST. Available online: https://www.nist.gov/pml/atomic-spectra-database (accessed on 5 November 2024).
62. Krstulović, N.; Labazan, I.; Miloević, S.; Cvelbar, U.; Vesel, A.; Mozeti, M. Optical Emission Spectroscopy Characterization of Oxygen Plasma during Treatment of a PET Foil. *J. Phys. D Appl. Phys.* **2006**, *39*, 3799. [CrossRef]
63. Cvelbar, U.; Krstulović, N.; Milošević, S.; Mozetič, M. Inductively Coupled RF Oxygen Plasma Characterization by Optical Emission Spectroscopy. *Vacuum* **2007**, *82*, 224–227. [CrossRef]
64. Jurov, A.; Popović, D.; Šrut Rakić, I.; Delač Marion, I.; Filipič, G.; Kovač, J.; Cvelbar, U.; Krstulović, N. Atmospheric Pressure Plasma Jet–Assisted Impregnation of Gold Nanoparticles into PVC Polymer for Various Applications. *Int. J. Adv. Manuf. Technol.* **2019**, *101*, 927–938. [CrossRef]
65. Qayyum, A.; Zeb, S.; Naveed, M.A.; Rehman, N.U.; Ghauri, S.A.; Zakaullah, M. Optical Emission Spectroscopy of Ar–N2 Mixture Plasma. *J. Quant. Spectrosc. Radiat. Transf.* **2007**, *107*, 361–371. [CrossRef]
66. Nutrizio, M.; Kljusurić, J.G.; Marijanović, Z.; Dubrović, I.; Viskić, M.; Mikolaj, E.; Chemat, F.; Jambrak, A.R. The Potential of High Voltage Discharges for Green Solvent Extraction of Bioactive Compounds and Aromas from Rosemary (*Rosmarinus officinalis* L.)—Computational Simulation and Experimental Methods. *Molecules* **2020**, *25*, 3711. [CrossRef] [PubMed]
67. Lamuela-Raventós, R.M. Folin–Ciocalteu Method for the Measurement of Total Phenolic Content and Antioxidant Capacity. In *Measurement of Antioxidant Activity and Capacity: Recent Trends and Applications*; John Wiley & Sons, Ltd.: Hoboken, NJ, USA, 2017; pp. 107–115. ISBN 9781119135388.
68. Yan, C.; Liu, H.; Lin, L. Simultaneous Determination of Vitexin and Isovitexin in Rat Plasma after Oral Administration of Santalum Album L. Leaves Extract by Liquid Chromatography Tandem Mass Spectrometry. *Biomed. Chromatogr.* **2013**, *27*, 228–232. [CrossRef]

69. Luo, L.; Kang, J.; Zhao, W.; Qi, Y.; Liang, S. Validated LC-MS/MS Method for Simultaneous Quantification of Seven Components of Naodesheng in Rat Serum after Oral Administration and Its Application to a Pharmacokinetic Study. *J. Pharm. Biomed. Anal.* **2019**, *174*, 1–7. [CrossRef] [PubMed]
70. Fu, Y.; Zu, Y.; Liu, W.; Zhang, L.; Tong, M.; Efferth, T.; Kong, Y.; Hou, C.; Chen, L. Determination of Vitexin and Isovitexin in Pigeonpea Using Ultrasonic Extraction Followed by LC-MS. *J. Sep. Sci.* **2008**, *31*, 268–275. [CrossRef]
71. Benzie, I.F.F.; Strain, J.J. The Ferric Reducing Ability of Plasma (FRAP) as a Measure of "Antioxidant Power": The FRAP Assay. *Anal. Biochem.* **1996**, *239*, 70–76. [CrossRef] [PubMed]
72. Brand-Williams, W.; Cuvelier, M.E.; Berset, C. Use of a Free Radical Method to Evaluate Antioxidant Activity. *LWT—Food Sci. Technol.* **1995**, *28*, 25–30. [CrossRef]
73. Nutrizio, M.; Maltar-Strmečki, N.; Chemat, F.; Duić, B.; Jambrak, A.R. High-Voltage Electrical Discharges in Green Extractions of Bioactives from Oregano Leaves (*Origanum vulgare* L.) Using Water and Ethanol as Green Solvents Assessed by Theoretical and Experimental Procedures. *Food Eng. Rev.* **2021**, *13*, 161–174. [CrossRef]
74. Koprivnjak, O.; Škevin, D.; Valić, S.; Majetić, V.; Petričević, S.; Ljubenkov, I. The Antioxidant Capacity and Oxidative Stability of Virgin Olive Oil Enriched with Phospholipids. *Food Chem.* **2008**, *111*, 121–126. [CrossRef]
75. Valcheva-Kuzmanova, S.; Blagović, B.; Valić, S. Electron Spin Resonance Measurement of Radical Scavenging Activity of Aronia Melanocarpa Fruit Juice. *Pharmacogn. Mag.* **2012**, *8*, 171–174. [CrossRef] [PubMed]

MDPI AG
Grosspeteranlage 5
4052 Basel
Switzerland
Tel.: +41 61 683 77 34

Molecules Editorial Office
E-mail: molecules@mdpi.com
www.mdpi.com/journal/molecules

www.ingramcontent.com/pod-product-compliance
Lightning Source LLC
LaVergne TN
LVHW070446061126
851858LV00011B/174